Lauren Hall Ruddell

Planet Goat Press

www.laurenhallruddell.com

planetgoatpress@gmail.com

Paperback Edition
ISBN 978-0-9984614-5-8

Contents

Prologue

Tiny green shards rested beautifully, if uneasily, across the hollow of her cream-colored throat. The emerald necklace she'd just received amplified the color of her own cat's eyes quite nicely. The auburn hair rested on snowy shoulders, overlapping onto the black velvet cocktail dress, and made the song "Jolene" come alive for her. Her favorite song of all time reminded her, of, well, her. Ivory skin and eyes of emerald green came from her genes, auburn hair from a bottle of expensive coloring. She watched the tiny jewels jump with each thrust of her jugular pulse. The tides of life, throbbing just below the surface of the skin, made her lustful. And satisfy that lust she would, in just an hour's time, with the sender of the necklace.

She estimated the necklace's value at just under $2,000. A nice piece and a nice peace offering, and one that might get him off the hook for next month's payment for her silence. She would just have to see how this evening's revelries went.

Tammy Splann valued only two things in life: money and sex. She gravitated toward men that had the former and wanted the latter. And given her looks, they were surprisingly easy to come by. And fortunately, all the best catches were married. Just as she was. It sure made her life easier than genuine romance with honorable intentions.

Thank God for married men! And thank God for marriage, that made the men in those situations bored and restless, longing for a bit of the strange, but too full of themselves to resort to prostitutes. Regardless of what she might think about romance, she knew that the sort of male ripe for the plucking was the delusional kind that had to imagine his paramour actually loved him, beyond hope, sanity, and common sense. At least for a while. Rich men could be so unpredictable, especially in the little backwater town of Peony Creek. She sure hoped that one day she would have the funds to blow this cow town, once and for all.

Her own husband Jack, an accountant, made a reasonable amount of money, but he was a tightwad, and uninclined to share the family wealth with her in the manner she felt she deserved. And working a nine-to-five job was just so boring! No, there were better ways to make a living that didn't even require selling drugs. And they were oh so much more entertaining than Jack. She believed that the phrase *"all work and no play makes*

Jack a dull boy" was invented specifically to describe her husband. So, a little on the side, here and there, with just the right sort, could pay dividends indefinitely, even after the affair was over. Especially after the affair was over.

It was with these cheery thoughts that she sidled from the inside of her vintage Mercedes SL and sashayed up the stairs to the designated hotel room her lover had reserved for this tryst. The stiletto heels, that she would keep on during their lovemaking and rake his back with, made a pleasingly Hollywood type of clack on the walkway tiles.

He liked it rough. He liked to choke her a bit. She obliged him because there was always an interesting gift or two. Sometimes even a check, made out to the charitable organization she was involved with but intended for her. The funds never made it to the charity, but did make it into an indexed fund, managed quite profitably by a married man from the old days, when she was still in high school. They hadn't seen each other in years, but she felt no need to remind him that his wife and the local authorities need never know what went on between himself and a minor. No indeed, and for as long as he so capably and profitably managed her portfolio, for none of the usual brokerage fees, she was content.

But this lover tonight, he needed regular squeezing of the financial sort, especially after he had ended their affair. Tammy, mildly disappointed but ready to move on, had agreed, for a price. Every single month. The gift that kept on giving. And at 22 years old, she had nearly amassed enough to settle comfortably on the East Coast among the sail set. Alimony from her husband of four years would be trivial, but income from several male 'admirers,' as she thought of her victims, would tide her over until she could score new prey. Which would be much easier in Martha's Vineyard than around these here parts, for sure.

With legs wrapped around her lover and his hands wrapped around her neck, she moaned with pleasure, both with the act itself and with the knowledge that her cell phone was set to record and doing so merrily in her purse on the adjacent nightstand. The hands closed tighter, and she gasped, as much for effect as anything else. Tighter, and the gasp was real. Tighter, and the gasp was no longer possible. Tighter still, and the lights faded, and it was good night, Jolene.

Late March was everything the locals disliked about springtime in the Rockies. Mud and goo everywhere. The trees were still mostly bare, with only emerging daffodils and tulips to bring a smile to one's lips and hope to one's heart. And this year one's heart needed hope, because a terrible pandemic had gripped the nation. But thanks to climate change, an unseasonable heatwave had dried the new fields at Bad Aspen Farm and CSA enough that Weston could get into them with the tractor. The new pasture would only be plowed and seeded at three quarter capacity, leaving a weedy patch of one acre close to the goat pen so they could enjoy fresh pasturage more of the year without bothering about seasonal rotation. One acre would be planted in herbs for the Merry Clary and the CSA shareholders, with basil featured heavily along with the more medicinal offerings. An acre for Saffron's Garden, of course, but that still left 14 acres to be managed for a baled hay crop, acres that had been neglected for years and had little left in the way of usable fodder. So, the sooner tilled, the better. After three days of temperatures in the 70s, the area was dry enough to keep the tractor from bogging down. The previous year's weeds would be turned under again and again as they emerged over the next few weeks. Only then would seed be laid down.

Weston wiped his brow on his elbow as he drove. His body, like everyone else's, was not ready for these temperatures yet, still being in winter metabolism mode. He noted that the workmen installing the elk fencing around Saffron's Garden were likewise sweating and that made him feel better. More manly. The pylons for the central gazebo were being installed as well. *Make construction while the sun shines*, he thought in a rare moment of literary cleverness. Then he turned his attention back to his rows. *Man*, he thought, *I sure hope Beth got a good price for this land, cause this is gonna take some work*, as his practiced eye noted some of the remains of last year's noxious weeds. His eyes had also spied some manmade objects that didn't belong in a high-quality pasture, such as rusty tin cans and a couple of discarded tires. He was on the lookout for anything that could foul the plow and put a strain on the already worn power take-off attachment. A large swatch of blue cloth just ahead poked up from the soil. He cursed and turned off the engine. Given what he found, the engine wasn't started again until later that day, long after the police had arrived.

Chapter 1

Head in hands, Laurelyn McNamara sat at her oak kitchen table, listening to the sound of the farm's small tractor being removed from the vicinity of the grisly discovery. Another body! This on her own property. She felt the chain of death and destruction might never end. This was the third death in one year's time. OK, maybe not strictly one year, since the body was plainly in a state of advanced decomposition.

Laurelyn had earlier glimpsed a white hand emerging from the soil. The face was thankfully still obscured by dirt, but a pale blue silk scarf was visible around the deceased's neck. Then Officer Mike Horvath had staked off the area with crime scene tape and all non-law enforcement personnel were asked to leave. Laurelyn and her business partner Beth were happy to comply. Weston, the part-time farmhand, less so. He stepped back a few feet and continued to crane his neck to see what he might see.

"You too, Weston, move on back now. Don't you have something better to do?"

"Not at the moment, not really," muttered the curious young cowboy, not picking up on the subtleties of Officer Horvath's suggestion.

"Then go find something. Now." Mike commanded.

"Ah come on man, I'm not contaminatin' the scene or nuthin' worse than it already is. Tractor probly saw to that. 'Sides, I found the body and called you guys to start with, so I'm a witness, right? Don't you need me close by for questioning?" Mike, hiding a lopsided grin at the follies of youth, replied, "Not this close. Go find something to do in the goat barn or somewhere."

Grousing to himself about stuck-up cops, he beat a slow, sullen retreat.

Saffron, the farm's permanent intern, was in the barn sterilizing milking equipment. She had neatly folded the freshly cleaned second-hand bath towels that were indispensable in kidding season. The five mature does that had been 'serviced' by an accommodating buck last November were enormous with young ready to make an appearance any second. Saffron was just sure they would all have triplets. But this was her first kidding experience, and she didn't know what a pregnant doe should look like. She just knew she felt so sorry for Belle, her favorite, whose udders were already enormous with colostrum. Her sides stuck out so far that you could set a plate on her back and it wouldn't slide off.

Autism notwithstanding, Saffron was an extraordinarily careful and thorough farm helper when it came to her beloved livestock. So when Weston burst into the milking room blowing and stamping soil all over the place, he received a rare admonishment.

"Out!" she shouted full in his face, before breaking off eye contact. Cursing, he left and went over to the tractor, the only being on the farm he felt really appreciated him at this point. He began changing the oil, even though it didn't really need it. His face was burning a little that it was Saffron who had chided him. She'd suffered from a huge crush on him only six months ago. He had renewed a romance with a local girl, Sally Handy, but at a subconscious level thought Saffron might always hold a soft place in her heart for him. Always nice to have a back-up plan, his reptile brain had reasoned. But Saffron had evidently healed entirely and although they were quite cordial and worked well as a farm team, Weston could tell that the flame had gone out, completely.

At the kitchen table, Beth did not sit with head in hands. Instead, she had initiated The Midnight Bark. Her first call was to Bernie Gonsalves, US Fish and Wildlife undercover agent extraordinaire. Bernie did not pick up. Beth called Bernie's mother at US Customs in the San Diego office and was told that her daughter was in the field in Costa Rica on a parrot smuggling sting and would be out of cell phone reach for another week. Then she dialed the last remaining number in the rotation, Officer Mike Horvath's wife, Sandy.

Sandy, who had heard the news from her husband an hour before, was finishing redoing the grout around the sink in the guest quarters and had just wiped the surface down when the phone rang. Seeing who it was, she picked it up and said without preamble, "Be right there!" She'd had plans in the vineyard cordoning up sagging vines in preparation for the new season, but that would have to wait. She knew if she got behind, she could always borrow Saffron for an afternoon. So, grabbing a bottle of their own vintage house white, she hustled out the door.

After putting most of the sleuthing band back together again, Beth got busy in the kitchen putting together her own version of Italian wedding soup. While Italian sausage bits coated in egg and panko sautéed in one skillet, she shredded a bulb of last year's fennel root to add to the stockpot that already contained the chicken broth and luxuriating onions, Swiss chard, and green garlic bits. These ingredients swirled in a tantalizing dance in and out of the cannellini beans. Soon the fragrance caused Laurelyn to sit up straight with a sigh of olfactory appreciation. Beth, who couldn't cook much of anything while her husband Ralph inhabited this earthly plane, was now coming into her own as a chef nearly as proficient as Laurelyn.

About the time Sandy was pulling into the driveway, the new medical examiner, Tabuache Ignacio, pulled in as well. Spying her stocky figure trudging down to the crime scene, Beth's eyes grew wide and she stepped away from the kitchen window. Entering the aromatic dining room through the back patio door that everyone used, Sandy glanced at Beth and then exclaimed, "Whoa, sister. You look like you seen a ghost!"

Beth cocked her head briefly in the direction of the retreating medical examiner. The said examiner had once not long ago had her arrested on suspicion of murder. Beth had neither forgotten nor forgiven. She retreated to the stove and made like a witch stirring her cauldron, all lowered head and intense concentration.

After a couple of minutes, Sandy intoned mordantly, "Come forth, she hath departed." She really wasn't gone, just out of sight for the moment, but that would have to do. With a relieved sigh, Beth rang the triangle outside of the back door to call the help in for lunch. She ladled a bowlful out for Saffron, the vegetarian, and then dumped the crackling Italian sausage bits into the broth, and set the whole steaming concoction on the adjacent dining table.

"Another death, can you believe it?" Sandy commented with a semi-gossipy, semi-serious tone.

"Actually, no, I really can't. What is this, some sort of curse?" Laurelyn lamented.

Beth, pouring wine for herself said, to everyone's surprise, "Well, maybe. There is a long history of hauntings on Long Hill up towards Ouray." At that moment the back door burst open, Weston entering first and Saffron second.

"Haunting, curses, now we're talkin!" Weston ladled enough soup into his bowl to sink the Titanic and plopped down.

"Well, there's bunches of stories about the Long Hill miner, who is seen quite regularly by everybody who lives up that way. But from what I can tell, he is just harmless and lonely, not at all malevolent. Oh and wait, there was a malevolent spirit reported in Hotkiss in 1969. Haunted a new pastor of the Church of Christ for three nights. The man left abruptly for Texas the morning of the fourth day, and never returned. So maybe this valley is haunted, but cursed? No way to know, but seems unlikely."

Laurelyn was staring at her business partner with open-mouthed surprise. The things you never know about people, she was musing. A paranormal buff! Weston burst out with, "Hot damn, oops, pardon me ladies, Saffron." The insult was not lost on Saffron and she managed to stick her tongue out at him before dropping her eyes to her soup again.

"Oh, I think there's malevolence afoot alright. MOOWHAAAHAAAHAA!" Sandy mimicked an old horror movie laugh. "But most probably of the evil human sort, with no further intervention from the unseen necessary."

Laurelyn, attempting to be the adult in the room, interjected, "OK, OK, I bet that's a given, but who is she, how did she die, and how did she come to be in my field?" Realizing that sounded self-serving, she amended it with, "I mean the new hayfield, Farmer Jeffries old acreage?"

All were quiet for a minute, when Sandy, in a moment of educated inspiration said, "I'll bet you if we knew 'how' she ended up there, we would also know who killed her and why. Or if we knew the 'when' of her demise, we might then be able to figure out the 'who.'

That fact could be very useful to us. Um, I mean to law enforcement investigators. I meant them," she muttered, her face suddenly pinker than the earlier raw sausage bits.

This faux pas got a smile from the stressed-out Laurelyn and another "Hot Damn," from Weston. "Bad Aspen Farm crime fighters are back in the saddle again. Another murder to tackle. This is awesome!"

In a second rare display of confrontation, Saffron said, "Please Weston, tone it down, that young woman was someone's child once upon a time." Weston, suddenly and belatedly remembering that the first body they had investigated had been that of her late boyfriend, colored deeply, upended the rest of the soup right down his throat, and excused himself from the table. A few minutes later, Saffron excused herself and claimed she needed to see to the turkeys.

The three main Bad Aspen Farm local crime solvers waited for exactly two minutes after the younger folk had left and then started in doing what they did best. Sleuthing. OK maybe they did farming and winemaking best, but sleuthing was a close runner-up. Since the outside professional help was incommunicado in Costa Rica, they would have to make do with their own skills. It was plain that Sandy would be the logical choice for chairwoman of the Bad Aspen murder-solving ad hoc committee, being as she had once had a couple of criminology courses before switching to agriculture. Being married to one of the town cops didn't hurt either.

Sandy loved her life in Peony Creek. The vineyard, winery, and B&B lodgings she and Mike had established kept her plenty busy. But it had been a long winter, and she had grown bored and restless. She had found herself looking back on her college years with nostalgia, recalling the excitement of learning and discovery. Her criminology classes had been absolutely fascinating, and in one of those courses, she had met her future husband. During their engagement, they had agreed that having two cops in the family might be one too many. Mike had already been in law enforcement for two years and was taking classes for promotional purposes. Sandy had always found agriculture interesting as well, especially viticulture, and so she agreed, with only a minor sigh, to switch majors.

She had never regretted it, and really didn't now. But she felt her brain might be turning to mush with all the busy work that innkeeping and vineyard management required. She knew she would feel very differently by October, but right now, a chance to help solve another murder struck her as just the thing to ease the ennui. Hence her slip of the lip.

Once her embarrassment subsided, she forged ahead with her detective musings.

"Why the middle of that field, I mean, parts of that site are actually visible from the road? Seems risky."

"I'm not sure, but I think there may have been an old building there that burned down a couple of years ago. Weston seemed to think so and blamed that for some of the trash he has been unearthing. I wonder if she was buried off to one side of it, or even inside, if the floor was dirt. The grave certainly seemed shallow to me. Like either the gravedigger was in a hurry or thought depth didn't matter. I'm betting that the body was buried next to the building on the side opposite the roadway because if it had burned down with her inside, the parts of the body I saw would have been burned as well. But that didn't seem to be the case, just normal, if revolting, decomposition," Laurelyn speculated.

"I concur," added Beth, before snatching up the lunch dishes and scurrying into the kitchen out of sight of the large windows that overlooked the two greenhouses, the goat barn and milking room, and most of the pasture. The ME was trudging back, along with the CIS photographer. The body was being loaded onto a gurney, having been first encased in a body bag. Beth had no idea whether Tabby would knock on the back door or not but was prepared to duck into the side room off the kitchen used by Saffron for crafts. She wasn't an overly shy person, but she was not as forthright as

Laurelyn and nowhere near as bold and gruff as Sandy. Sandy's 5-foot, 11-inch frame, and Viking looks and manner made her a force to be reckoned with if she was upset. Only Agent Gonsalves was her equal in that department. Beth was made of softer materials.

Sure enough, a knock on the door. Beth was already out of sight but listening from behind the craft room door.

"Hello Ms. McNamara, we have completed our initial investigation of the body and the grave in situ, and now we are heading back to the morgue, where I will complete my examination of the body. I will need to be in contact with you about some details concerning the pasture's uses in the past, the discovery of the body, and other items as more information on the victim emerge. You can continue plowing your field if you need to, but please ask your man to try to avoid that immediate area for at least another week. I'm assuming you have no idea who this could be? No, I thought not. But if anyone local thinks they know, please have them call me. And tell Beth Archuleta I saw her in there. SORRY BETH, I was just doing my duty as I saw it. Have a good day and I will be in touch." With a wry smile, Tabby slouched off to her SUV.

"Well, that one certainly doesn't miss much. I bet she knows who the victim is by the end of the day. She already seems to think it was foul play, and not suicide covered up by an ashamed family member. Wounds on the body? Strangulation? I'll see what I can get out of Mike, confidentially, of course," Sandy assured them with a wink.

"Oh, of course, of course, your lips are sealed and we would have to torture you for the information," Laurelyn snarked.

The sound of the tractor starting back up interrupted their confabulation.

"Need me for anything while I'm here? Help in the greenhouse for a bit? OK, will do and then I need to borrow Saffron for a little bit. I have vines in desperate need of restraint, and with this new crime to attend to, Mike is going to be working overtime for the next couple of days, I'm sure. Let's hit it, Beth, since the wicked witch is long gone. I'll collect Saffron in about an hour," Sandy announced, plainly relishing being able to take charge. Laurelyn smiled and made shooing motions with her hands.

With them gone, she took several deep breaths, put on some soothing new-age music, and sat down at her computer to update the farm's website. Community Supported Agriculture shares for members would start in just over five weeks, and God only knew when the first baby goats would make an appearance and if the birthing would be without trauma. She hated to have to be indoors on such a lovely day. But knowing that someone had died out there, on her land, made her happy to be able to cocoon herself in her small bedroom/office for a while.

Once she felt she had made significant progress, updating dates and deliverables for this year's shares, she sat back with a sigh and texted her boyfriend, George Trojan. She hadn't reached out to him, knowing that he had a mare in foal who was plainly in the last stages of gestation. Besides, his cell phone service up on the high mesa where his own cattle ranch was situated was sketchy. Her call would be detailed and complicated, so she just texted him that there was some interesting news and to please call her when he got a chance. She knew that he would eventually, but for now, he was probably instructing his son William in the fine art of livestock midwifery. A knock on the door and a quick wave from Sandy indicated that they were done and that she was shanghaiing Saffron for a couple of hours. Laurelyn quickly grabbed Saffron's coat and work gloves, which she thrust at the departing pair.

Laurelyn had become a surrogate mother to the sensitive and damaged Saffron. Saffron's mother, a former stripper, had never been able to tend to the child's needs properly, nor protect her from an abusive stepfather. Now that she had stability, she had blossomed into an indispensable helper who earned her own keep many times over. From a withdrawn airhead to a sensible and artistic young adult, the transformation had been amazing. Truth be told, it had taken a whole village to make that happen, but happen it had, with help from Beth, Sandy, and even George. Such was the power of community. So, murders aside, Peony Creek was a pretty awesome part of the planet, in Laurelyn's opinion. She finished her website modifications and grabbed her coat to head back out for her own chores. And some speculations.

She pondered the paradox of Peony Creek as she meandered down to the pasture. She gazed into the now-empty grave and wondered who the woman was. A local? A tourist? A wife, mother, daughter, lover, friend? What roles had she played in a life ended early? Was she a good person, or a bad one?

Weston noted her silent reverie and turned off the tractor. He came over and stood beside her, unspeaking, sensing she was offering kind thoughts for the dead woman. Out of the corner of her eye, she noted that he had clasped his hands together in front of his lower body, as if at a funeral. She found the gesture oddly endearing. So typical of her feelings about Weston. One minute his boorishness made her want to scream, and then the next he did something that made her want to pinch his cheek or ruffle his hair. Finally, she turned to him and asked, "You don't know who she might have been, do you?"

"Oh gosh, no. People don't just disappear that often around here. A summer tourist, maybe? There was a hippie girl that went missing five years ago and a local gal that ran off unexpectedly about three years ago, but those are the only folks I can recall. As the cops walked past, I heard something about a 'hyena bone.' I thought that was a weird thing to say. Maybe the dead woman was a biologist or something."

Laurelyn perked up and asked, "Could it have been hyoid bone?"

"Yeah, I think that was it, a broken hyena, I mean hyoid bone."

"Huh, then that's why the ME is going with murder so quickly upon cursory examination. Those are little bones in the neck that are always broken when a person is manually strangled. I wonder if she was strangled with that scarf that was around her neck? Well, none of our business, I suppose. So, I guess we both better get back to work. I will be in the goat barn if you need me."

She had been able to sense Weston's rising interest and wanted to derail it before it got out of hand. Hypocritically, she allowed her own interest to rise, and as she strode toward the goat barn, she texted Sandy that it seemed likely the victim had been strangled.

A shriek from the goat barn caused her to cram her phone back in her pant's pocket and race to the barn. Weston came pounding after her. Unless there was another murder in progress, kidding time was at hand.

Chapter 2

It was Belle, she was down on all fours, and leaning to one side. Her vulva was puffed outward and the tips of two tiny feet could be seen. Her water had burst some time ago. Laurelyn grabbed four of the folded towels, the goat birthing kit and directed Weston to get Saffron. After that, he needed to make a thin oatmeal broth on the kitchen stove for the new mother and then get his butt back down to the barn.

The presentation seemed normal, at least for the first baby. With a mighty groan and push, the little hooves cleared the birth canal, with the tiny muzzle resting on top of them. A second push and the shoulders cleared.

Laurelyn waited for the third push, grasping the baby behind the shoulders. With the third contraction, she gently pulled the baby all the way out and onto the clean towel. No sense in asking the doe to work harder at birth than necessary, she had learned. With a second towel, she quickly wiped all of the birth fluids away from the baby's mouth and nose. She vigorously rubbed the baby down, including the insides of the pendulous ears. Only then did she lift the newborn up to dip its navel in iodine and check the sex. A girl! Hallelujah, because if this little one was the heavy milker and sweet-tempered angel that her mother was, she would be a valuable asset to the farm. Realizing that Saffron had been returned to the farm, and was now standing a few feet away, eyes wide in awe and tears running down her face, Laurelyn gleefully announced the little girl's name.

"Bad Aspen Farm's Arabella, in honor of her mother." Saffron looked up and laughed out loud with delight and relief. Her joy didn't last long, as another loud groan announced the arrival of a second baby.

One foot, and a contraction, and still only one foot. Crap, it was going to be a breech birth. Laurelyn quickly grabbed alcohol and nail clippers out of the birthing kit. She clipped all of her nails as short as she could get them and dowsed her arms to the elbow with the alcohol. KY jelly smeared on the back of her right hand, and in she went. Belle hollered with pain, while Laurelyn felt for the backward foot. She found the head. It was forward. One less problem. Then she found the backward foot and inch by inch pulled it forward. With a finger crooked behind the tiny knee, she pulled. Then with finger behind the tiny ankle, she pulled. Finally, the tip of the small hoof presented itself next to its mate. With Belle in such pain, Laurelyn did not wait for further contractions, and with one hand behind the tiny head and the other grasping both hooves, she slowly but firmly pulled the baby out. A gush of pent-up blood followed. Laurelyn heard a thump behind her but paid it no mind since she had to assess the quality of the blood. Dark with birth fluids, so no hemorrhaging, at least not yet. She wiped the baby down. Another girl! Smaller than the first but that turned out to be a good thing. After she had placed both babies in front of Belle's face so she could lick them and get acquainted, she stood with her balled fist in the small of her aching back. She turned to see what Saffron thought of all this and whether the warm oatmeal broth was ready for the now visibly relaxing but weary mother.

The thump she had heard became plain. Saffron had fainted. Weston was holding her head in his lap.

"Did you see her go down? Did she hit her head on anything," she barked at him frantically.

"No, I had just put down the bucket cause I saw her sway and I know how city girls are. She sorta melted straight down. I couldn't catch her all the way, but I was able to grab her elbow before her head hit the ground. At least hit the ground very hard, anyway."

"Fast work, and thanks. I don't see any blood, but I bet she has a headache. Ooops, hey, there you are my dear! You gave us all a scare. Can you sit a bit? Not fast, just take your time. That's a girl. Now, did you hit your head? Oh, a little, right there. Yes, I think you may get a goose egg on your forehead. Goose egg, that means a lump. Weston, can you get her an ice pack from the milking room?"

Another groan, and not from Saffron. Triplets!! Saffron had been right all along. Two hooves, an enormous head, and the little buckling nearly shot out of the womb. Laurelyn barely had time to scramble over with the last clean towel on hand.

"Gotcha! Weston, another towel and give Belle her oat broth." Laurelyn knew her abruptness was rude, but this day had not gone the way anyone had imagined, and the stress was taking its toll. Finally, Belle got to her feet and began to lick all three, crooning and chortling. The little ones bleeped their first sounds, with the first baby already trying to stand. The bleeding was not significant, so no tearing of the vaginal passage had occurred. Laurelyn would give Belle an antibiotic tomorrow anyway for good measure, but not today, since she didn't want it in the babies' first colostrum.

"Weston, time for a break for us all. Three folding chairs and beers all around, whaddya say?" More than happy to drink and rest while on the clock, he sped off to the greenhouse where the chairs were kept. Even Saffron, who rarely drank, had her own beer, mainly so she could toast to the new babies, but also because the day had been awfully hard on her as well. Laurelyn had already decided to register and sell the smaller of the two does.

The buckling would be castrated in a few days to make him into a wether. In a month he would be introduced to wearing a collar, the month after to lead on a leash, and the month after to follow on command. At five months he would wear a lightweight dog harness, then dog panniers. At six months old, he would be sold as a well-started pack goat, along with any other males born this season.

Weston and Saffron finished their beers and went back to their jobs. Laurelyn stayed. She wanted to make sure all three babies had found the teats and had full bellies before the chill of the evening came on. Which was not far away, this still being March after all. She removed the towels and freshened the pine shavings. After hanging the towels outside and hosing them off, she popped back inside and found the little boy nursing greedily. She knew the other two had supped and were already asleep in the corner. Satisfied, she turned up the barn heat a little and turned the sound up on the barn monitor. With four more girls to go, the odds of burning the midnight oil were excellent. She would sleep in her clothes tonight and every night until the last doe had delivered.

As she plodded up to the house, relieved and pleased but bone-weary, she contemplated new life arriving upon the heels of death discovered. A strange day for sure. The circle of life was a constant drama, it seemed. Circle of life thoughts reminded her of George. Had the foal been born? She checked her text messages. Nothing there yet. That was because the man himself was striding down the path toward her. Grinning, she threw herself into his outstretched arms.

"George!!" a shrill cry quavered over the farmstead. But not from Laurelyn. No, it was Saffron, almost fully recovered and rushing to her favorite Dutch uncle, too excited by the day's news to let the lovers have a moment's peace. Missing the wry smiles and rolling eyes of the pair, she grasped George's hand and dragged him off to help her close in the turkeys and chickens for the night. With a resigned shrug, Laurelyn granted them permission and went kitchenward to heat up the wedding soup for him and make grilled cheese sandwiches for them all. Once Saffron had gone to bed, they could head to the goat barn for a late-night snog and baby goat watching. And conversation, in between the snogging. And boy, did she have a lot to tell.

Chapter 3

Just about the time that Laurelyn, Weston, and Saffron were celebrating the new triplets' successful birth, Tabuache Ignacio was contemplating an untimely death. She stroked her chin as she gazed at the victim's fully exposed corpse. She had judged correctly that given the excellent care the deceased's teeth displayed, dental records with an ID would be forthcoming shortly. The enamel on the teeth was far too thin for a person her age, indicating constant teeth whitening procedures. This indicated that the victim was both vain and wealthy.

She was not wrong, and the identification was returned to her within two hours of the request. Tammy

Splann, ne Bedecker, had seen a local dentist until her disappearance three years ago. The dentist assumed that she had moved away and failed to have her records transferred. He had found it odd since her husband still lived in the area. But he just accepted that this was the way folks were and gave it no further thought. On the phone with Tabby, he confirmed that she had grown up in Peony Creek and graduated from the local high school. He also confirmed that she had been a beautiful woman with a bad reputation, bad as in slutty. She had always seemed very pleasant to him, and gave him plenty of business, having her teeth whitened every six months religiously. He was sorry to hear that she was dead but wasn't terribly surprised. She'd had a reputation for a high-risk lifestyle. He had been surprised that she had survived as long as she had, and not killed herself sooner the way she drove that Mercedes of hers. Perhaps Tabby should talk to her husband, Jack Splann, for more information? Tabby noncommittally thanked him for his help and hung up. Returning to the body, she continued her external examination.

The signs of strangulation were too broad and non-localized for the scarf to have been the murder weapon. No, definitely hands involved, the hands of a large and strong attacker. The scarf had therefore been added later. But why? Was it a comment on their relationship? Had it been a favorite article of clothing? A sign of remorse by the killer? Before making any incisions, she dictated these thoughts to herself and then set to work. Before making the first cut, however, she texted the Peony Creek Police Station to advise the cop shop that a chat with the former husband was now a critical part of this investigation.

The morning following the grisly pasture discovery, up at Mesa Splendor vineyard and winery Mike Horvath got a call from his boss as he was tugging on his uniform pants.

"Yeah," he spoke at the phone, "Yeah, OK, will do." His wife, Sandy, said with no preamble and even less shame, "Was that about the murder?"

"Yes dear."

"Well, what did he say?"

"Stuff."

"Ah come on, don't 'stuff' me."

"You didn't mind last night."

She threw a pillow at his face.

"No really, what did he say?"

"Things."

"Michael Horvath, if you ever want to have sex again, you'll tell me what he wanted."

"As you wish, my lady." He had been intending to tell her all along but loved tormenting her. He satisfied her curiosity by letting her know that he would be stopping by the office of Jack Splann this afternoon. Her eyes grew huge.

"No, no way. The victim was Tammy Splann?" she exhaled the words, jumping to the correct conclusion.

"My lips are welded shut. See you tonight." And he sprinted out the door, but not quick enough to avoid the next pillow that smashed him between the shoulder blades. The minute he was out the door, she dialed

Laurelyn. She was surprised when Saffron answered the phone. Laurelyn was usually up cooking breakfast by now, but not outside quite yet.

"She's still asleep," the small ghostly voice informed.

"You're kidding! Is everything OK?"

"George was here late, and then there were more baby goats. She has only been in bed three hours."

"Crap, of all the days. Are you OK with the morning chores?"

"Well…"

"Be right there," she said for the second morning in a row. She knew that Saffron would be perfectly fine with all chores, but needed a comforting presence in the form of an older adult. She texted Beth, requesting early assistance in the greenhouses, and then started getting dressed for a chilly morning. She grabbed pruning shears, and in the false-dawn, clipped off the last of the straggling vines that had been left from the day before, when it had started to get too dark for her to work. That done, she gathered up the vines to give to the new moms as a treat and headed down the small steep road from the mesa to the valley floor. The tires crunched on crystalline gravel, letting her know that this day would be chillier yesterday since the road was well-frozen. She hoped the baby goats were all OK. Having a desire for fast info, she called Weston about what he remembered about Tammy and the old building.

She pulled into the small farm store parking lot, grabbed the vines, and headed for the goat barn. Entering, she spied Saffron already there with two baby goats tucked into her enormous down jacket.

"They're cold. They're shivering. What do I do, Sandy?"

"Here, give me one. I will stay here with one while you take the other one up to the house. Woodstove going? Good, then get that one warmed up for half an hour and then bring it back down here to its mom.

"Her mom. Thanks."

The new mother, a nervous two-year-old lacking Belle's phlegmatic life perspective, crooned and fussed, sticking her nose inside Beth's jacket to reassure herself her child was safe and well. She hadn't yet realized that one was missing, but she would, at some point. Sandy wanted to distract her with some vines, but when she stood up, the doe fussed and cried. So back down in the cold straw she sat. At least it was clean, having been replaced at 2:00 AM by George and Laurelyn after the birthing. This birth had been textbook, which came as a relief to everyone.

Up at the house, Laurelyn was finally up, if bleary, and making coffee and cocoa. Quick biscuits and gravy with elk sausage were being whipped up and the fireplace was roaring. Towels from birthing were in the washing machine in the little side room laundry but hadn't been dried yet. Saffron entered and plopped down by the hearth with the little doeling. She put her hand under her jacket and pulled it back out with a "yeaouuuuuuuu." Fingers covered in sticky black tar were held up for Laurelyn's inspection.

"Meconium," she pronounced, and went to that bathroom to get her own personal bath towel to clean the baby goat and Saffron alike. The coat would definitely need some laundering as well. After meconium would come the first real bowel movement from nourishment taken outside the womb, but the first poop was always nasty.

After all hands were sterilized, she handed Saffron a plate of biscuits dripping in butter and honey, the way she liked it. She decided to wait on her own breakfast till Sandy came up to the house. A half an hour by the fireplace did the trick for the chilly kid. When Saffron had finished her biscuits and sat back down in front of the fire, the baby goat began punching her neck. Plainly hungry, she was looking for a teat, which was not to be found under a human chin. The little girl was hustled back down to the barn and placed under the mom, who, puzzled at first, became content when the doeling began nursing.

Sandy, chilly now herself, was more than happy to remove herself and the little buckling to the warm kitchen. The sun was now fully up and thankfully the temperature in the barn already improving.

The buckling was not as cold as his sister, so Sandy was able to eat at the table with one hand shoveling breakfast into her mouth, and the other hand cuddling the baby boy next to her heart. He must have had a bigger meal than his sister before his first nap this side of the womb, because after 10 minutes, he struggled and let out a loud "BAAAP." Sandy put him down, and he proceeded to make his first tentative jumps. He managed three before running out of steam. Laurelyn wrapped him tightly in a kitchen towel and placed him in her lap so Sandy could finish her meal.

After letting out the turkeys and feeding the chickens, Saffron appeared and took the drowsy boy back to his mom. It would be best if he could nurse again before his next nap. Saffron was more than happy to play nanny to the nanny and her youngins. Despite her thinness, when properly dressed Saffron was very impervious to cold. She had proven to be a godsend in so many ways at so many times, and this was one of those.

While clearing up the breakfast dishes, the women spied Beth heading to the greenhouses. Although Beth had little interest in the animals above enjoying their antics, she had proven to be a genius at gardening and greenhouse management. The chiltibarn, the greenhouse containing Laurelyn's patented peppers, was still the head farmer's domain, but the herb and vegetable greenhouse had been relegated to Beth and Saffron, and the results spoke for themselves.

For a few precious moments, Laurelyn had time to gossip before goat barn duties. Helping herself to her third mug of espresso, she sat at the dining table with Sandy.

"So, guess what?" Sandy had learned to torment listeners from her husband.

"What?"

"Guess!"

"Goddammit, not now. What?!!"

"Fine be that way. I know who the victim is."

"Shut the front door! How? When? Who?"

"Mike, this morning, Tammy Splann, and Colonel Mustard in the drawing-room with the candlestick."

Laurelyn stuck her tongue at her, then thought for a moment. "Isn't that the young woman Weston mentioned disappeared about three years ago? She's a local, right?"

"Correcto, and Mike is going to interview the husband today."

"Crikey, that was fast. I know they say it's always the husband, but after our experience with Beth, I hope all concerned can keep an open mind. Jack Splann, that accountant in town, right?"

"Yep, and that fellar is in for a bad day."

"I'll say. But wouldn't it be seriously lame for him to kill his wife and bury her in a shallow grave on private land when as a local he must know a million better places in the wilderness to do that?"

"Seems like a Darwinian low-fitness move. But who knows? But what we do know is who, and because of who, we know when...three years ago. Old man Jeffries old shed was still up then, but not for long. Rumor, that is Weston, has it that one of Jeffries' nephews used to go there to smoke, something his parents discouraged since the kid was only 12. But something happened, the kid caught the shed on fire and beat a hasty retreat. It only came out days later who the culprit was and why. Rumor also has it that he couldn't sit down for a week. Ah, the good old days, when you could beat children for misbehaving," quipped the childless Sandy.

"But we don't know who killed her, who buried her and why there?"

"True that, and the plot thickens. I'm going back outside to get another baby goat fix. It's hot in here. Coming?"

"Yes, mam, I wouldn't want to slack off on running my own farm, now would I?" Sandy grabbed her favorite assault weapon from the back of her chair, and with recent practiced expertise, threw it at Lor. The pillow hitting her between her shoulder blades made her yelp. Equity thus established between mature individuals, they headed out together.

On a whim, as she walked Sandy texted Weston again. He would be helping George this morning with stable chores since George and his son William had gotten behind in all the murder hoo-haa. Her message said simply, '*Bad Aspen this afternoon*'? Then she trotted toward the warming goat barn from which the sound of tiny bleats was wafting.

As she milked out a little colostrum from the huge uddered Belle, to save for baby goat emergencies, Laurelyn had a chance to think while Sandy played with the triplets. What was the motive? Had the dead woman known something about someone that put her in danger? Had she done something to incite a friend, family member, or stranger to violence? That motive would be incredibly important in finding the killer.

Then she turned her thoughts to access to the burial site. The parcel had always been used as pasture, and the only structure on it that old equipment shed. Barely a double track, an incredibly pitted dirt road had once served the little shed and was the property's only access from the main road. That was back in the day, for now the pasture was fully fenced and the road plowed under. Access to the pasture was through Bad Aspen Farm grounds only. But Laurelyn remembered how rutted the road had been when it still existed. Only a four-wheel-drive vehicle would have been able to attempt it, and only in dry weather. Had the murder occurred right there, on the site? That was a very unpleasant thought. But it did mean that the body dumper was not a tourist cruising through the area in a sedan. She wondered what Jack Splann drove and what his wife Tammy drove.

"Sandy, did you ever run into Tammy Splann? Were you guys living here before she disappeared?"

"Yeah, actually we were. I didn't know her, but when she pulled up to Dawn's Market in a fire engine red Mercedes and flounced into the store in stiletto heels, well, she got my attention. I was chatting, gossiping really, with Margaret Stanley in the parking lot and asked her who that was, knowing she would know. She did, and pronounced her to be Tammy Splann, the town's high-class slut. 'And oh my gosh, her poor husband, such a nice man, I don't how he puts up with it...' and so forth."

"So she was driving a sports car. What does her husband drive?"

"How would I, oh wait, I do know what he drives. He leases a new Lexus every two years. Makes Mike so mad he could spit. It's none of his business of course, but conspicuous consumption really plugs him in, and whenever we are going somewhere and he sees Jack in a new car, he makes a nasty comment. What makes you ask?"

"Just pondering the burial site a little further. Did you ever notice a gate across the road to that old shed? No, huh, so free access, in the right vehicle. I wonder if the police will think to ask Jack about his transportation options?"

"Oh, I see, still trying to avoid another Beth, are we? Very civic-minded of you." With that, she held up her pointer finger to indicate waiting a moment. She texted Mike. '*Remember Jack drives Lexus, could be important.*'

"They will now. If you're all done here maybe we should go back out into the field. We may notice something they didn't. Them being the Keystone Cops and all." Sandy had a very high opinion of her husband's professionalism, along with a certain young public defender and the ME, but that concluded the list of approved law enforcement personnel.

Laurelyn smiled and agreed, just as soon as Sandy gave all of the goats their grain, and the pack wethers got their hay and minerals.

Gazing at the empty grave was a real downer. Sandy broke it off first and began to walk in concentric circles around the site. All she found was part of an old plastic jug. Returning to the graveside, she saw that Laurelyn was frowning and squinting in intense concentration.

"Does the bottom of the pit seem discolored to you?"

"I guess a little, maybe somewhat greenish. Why?

"I wonder what that is?" Laurelyn whispered to herself, not answering the question. "If I climb down in there, can you give me a hand back out?"

"Why would you want to? And sure, if I feel like it."

Ignoring the snark, Laurelyn patted a jacket pocket. Yeah, she still had an old, sterilized margarine container and lid that she hadn't used for colostrum that morning. The grave was so shallow that the ground level was no higher than mid-thigh, which she determined only after she had jumped in. Taking the container, she scooped up some of the puddled dirt where it was the most greenish. Sniffing at it, she nodded her head, as if affirming something. She capped the container tightly and handed it to Sandy. Sandy used her Vikingish strength to easily haul Laurelyn back out of the hole.

"So, Poirot, what have zee little grey cells been telling you?"

"If I'm right about what's in this tub, this may not have started out as a burial site and may not have been dug by the killer at all. Old man Jeffries has a really big hayfield somewhere else in the valley, doesn't he?"

"I don't know, but Beth might since she did the negotiating and purchase of this field from him. Maybe it came up in conversation."

"Good thought, let's go ask her."

"Let's do, then I have to scoot into Delta for groceries and more grout."

Beth had finished removing the double cloths over the winter greens that would not be needing them anymore until next December. She would leave hoop removal for Saffron. The system of double greenhouse fabric covering half-round hoops within the greenhouse itself had ensured that the farm had salads all year long, a modification to the farm operation instigated by Saffron and supported by Laurelyn and Beth. Beth had just finished rolling and tying the cloth for storage when Laurelyn and Sandy entered.

Laurelyn stopped, awed for a minute by the lush growth under the canopy. She'd heard it had gone well but seeing is believing. She would make this system a permanent feature every winter, and maybe even expand it.

"Impressive, most impressive. But I got a question for you. When you were working with Mr. Jeffries on the land purchase, did he mention whether he had more fields or not?"

"Yeah, actually he did, trying to impress me. I think he may have been flirting a little, being a lifelong bachelor and all. He has 148 acres down the valley toward Hotkiss in the Midway area. He bragged about its current value, should he decide he wanted to subdivide it. Why, you looking to buy more land?" She added this remark with a twinkle that belied the query because she already knew the answer would be 'no way.'

"It may have a bearing on the body in the grave. Which may not be a grave at all. Here, now that we are out of the breeze, Sandy smell this. You too Beth. Does it smell sweet to you guys?"

They both had to take several tries, but eventually agreed that it did, just a tiny bit. Beth started to stick her finger in it.

"I wouldn't. I think this is anti-freeze. Old and decayed into mostly its constituent parts, but still full of heavy metals. If it is, then it was poured into that pit in such enormous quantities that it still retains its color, which it shouldn't after all this time. I need to have this confirmed by a lab, but I'll bet you anything that's what's down there. I'm guessing that the pit was originally dug to illegally dispose of toxic farm chemical waste. Jeffries has a reputation around town for being so cheap, he makes every penny scream for mercy. It wouldn't surprise me if he decided that proper disposal was too expensive and was dumping all manner of substances down in that pit."

Beth, looking increasingly horrified, said, "Does that mean the land is worthless for crops?"

"No, not at all if I remember my ag chemistry correctly. At worst, depending on the quantities, that pit may need to be excavated and refilled with clean soil. If it's only anti-freeze, at that depth it's probably OK, but if there are other things? Well, let's wait and see what the soil guys in Delta make of this before we get too upset."

"So, if I follow you," interjected Sandy, "you think the killer knew the pit was there next to the equipment shed. He knew the land was for sale and wasn't being maintained anymore. He knew it wasn't gated and locked. He lured Tammy there on some pretext, or killed her elsewhere and brought her body in an SUV of some kind, and dumped her there, then shoveled over the pit. You know, it strikes me that the killer must have been in a tremendous hurry. With a little more top cover, and rolling back and forth over the pit with the vehicle to break up the corpse and smash it more deeply into the soil, the frost-heaving might never have worked the body to the surface to the point that the blue scarf emerged above ground."

Laurelyn and Beth both stared at her in horror of her depiction. Then they nodded in concurrence.

"Certainly what a professional would have done. But a professional would not have deposited the body in such a shallow pit to start with. Animals could have dug the body up long ago. So a scared or hurried amateur, not a professional killer. If any of this is true, it makes it seem much more like a crime of passion than a mugging or rape gone wrong. And with those cheery thoughts, I bid you adieu. Gotta keep a hungry cop fed. You know how cranky hungry cops can get."

"Indeed, but if you're going down valley anyway, could you drop this off at the soil lab on 5th street in Delta?"

Watching Sandy's departing back, Beth said, "You mind if I work in the chiltibarn a little with you? I don't feel like being alone right now."

"Of course, and you can do the record-keeping as I plant the new seeds for the year. I see Saffron over there raking out the turkey pen, so unless there are more baby goat births, let's plant those peppers and ponder their propensities."

Beth smiled at Laurelyn's weak alliteration. Even though awkward, it served its intended purpose to lighten the mood. The task needed to be done, and the greenhouse temperatures were perfect more than two weeks early. Besides, planting over 50 pots was fairly mindless, and would give Laurelyn a menial task that would leave her mind free to wander.

She didn't know for sure that this toxic waste aspect of the crime scene had been overlooked, but she would bet any takers $20 it hadn't been noted yet. Still, if the body had been in contact with that supersaturated soil, green dye may have permeated the skin, and Tabby would identify the source of the discoloration. Laurelyn wondered if skin coloring was possible after death.

Surely it was, or how would people dye leather? Shocked at the gross turn of her musings, she took several deep breaths. Her mind cleared, and she deliberately thought of something else.

However, after answering a few questions for Beth, her mind slithered back into solving the crime while keeping her farm running. Should she bring the soil to the attention of the authorities? If she did, would it create more headaches for farm operation scheduling? Was that selfish and irresponsible of her, or would revealing the results of the sample equate to poking the bureaucratic bear? What would Bernie do?

She missed her friend's counsel on these matters. But in the absence of such counsel, she decided to do nothing until tomorrow, if then. That decided, she asked Beth to keep the information to herself and not even

Saffron and George were to know. She texted Sandy with the same request. Then, rounding up Saffron, the three women went up to the house for lunch. As they sat down to Laurelyn's famously decadent potato cheese soup and sourdough bread croquettes, her phone chimed a text. Weston had gotten back to Sandy. Sandy was relaying that Weston's girlfriend, proprietor of the local herbal shop and one of the farm's main customers, had known Tammy Splann quite well. They had gone to high school together. Weston would collect the yearbook that Sally had offered and bring it with him this afternoon when he came for half-day chores.

"Oh wow," Laurelyn exclaimed, and relayed the gist of the text. She then texted Weston and asked if Sally could send along a spring herb order request as well. They more or less knew what she would be wanting, but not how much. Beth offered to stay and take charge of the order, if Laurelyn wanted to go down and inspect the new gazebo in Saffron's Garden since the crew had left when the police arrived and had just now returned. That decided, a bowl of soup with toast was set on a hot plate for Weston. Laurelyn headed down and Saffron went with her. After all, it was her dedicated garden, courtesy of Beth. As they walked, the turkey boys, Tom and Jerry, strutted after them all the way to the construction site. Which cracked up the Hispanic construction crew a great deal. It appeared that the gazebo would be completed by nightfall. The fencing would be finished the next day. With a sigh that something was going right, the women and turkeys marched back up the slight slope. A holler issued from the goat barn.

"Oh my, here we go. Saffron dear, can you run and get all of the dried towels out of the dryer and bring them down to me. And when Weston arrives, send him down, but you yourself need not hang out. No fainting needed to end this day with, and no need to add another goose-egg to your damaged brow. No, I am so not mad, because after the birth is over, you get to do the clean-up. Listen for my cue in the barn monitor and come down when all of the gross stuff is over."

Embarrassed but relieved, Saffron sprinted for the dryer. With hands so full of unfolded towels she could barely keep from tripping, she sprinted back down the slope to the barn. She heard but paid no heed to the car pulling into the parking lot many yards behind her. Handing four towels to Laurelyn, she folded the other six and put them on the shelf where they belonged. She handed her the birthing kit and then spotting blood and a tiny hoof, she fled to the house.

It was on this flight that she nearly ran into Weston and Sally. Sally had decided some explanation was needed on her herb order, and so had closed for lunch and come to the farm herself. Saffron stammered out the goat news and then kept running as commanded. Weston took Sally by the hand, assuring her she was in for a real treat.

Although Sally was a country girl, she was not that kind of a country girl and after a few minutes, excused herself to confer with Beth on the timing of deliveries for the goods ordered. Weston and Laurelyn assisted with the fairly straightforward birth of boy/girl twins. Laurelyn saw that the little boy was hyperextending his back legs, which would need to be attended to with splints in the next few hours. Other than that, all was good. Weston wandered over to check out the construction, and then went up to the house to tease Saffron and collect Sally to take her back to her shop.

When Laurelyn later reached the house, she was greeted by the sight of Sally's yearbook spread open on the dining table, and not the herb, vegetable, and flower order she had expected. The picture of interest was of a young woman of exceptional looks, who would be a stunningly gorgeous adult one day. As Beth and Laurelyn looked on, Sally showed even more pictures...Tammy as homecoming queen junior year, then again senior year. Tammy as cheerleader, Tammy as varsity tennis star. Tammy as prom director. Tammy as yearbook assistant. Yes, it was a small high school, but Laurelyn wondered how one teenage girl could loom so large in the school's psyche. Sally explained that her parents were quite well off, having a majority share in a local coal mine employing 200 people. OK, yeah, wealth certainly matters, but was she super sweet, super smart? No, Sally assured her. She was average in every way except looks and charm. Charm directed at men and boys was a specialty.

Intrigued, Laurelyn asked, "Sally, do you mind if we keep this for a while? We'll take good care of it, I promise."

"Oh sure, keep it as long as you like. I haven't looked at it in years. Gotta run now. Can I come up in a couple of days and buy some greens? I know it's early, but Weston spilled the beans and said the greenhouse is already rockin'."

“Absolutely, and before you go, all the yucky parts of birthing are done and the baby goats in the barn are all super cute, so pop down there for a minute if you care to take a peek.”

“Was that wise?” Beth asked after Sally left, referring to the recent love triangle. Laurelyn having closely observed the interactions between Weston and Saffron, assured her that it would be fine, that all was well in that department.

“Oh cool, well, I think I will take this order home for a bit. It’s busy enough here that it’s hard to concentrate.”

“Oh you bet, can I join you? Just kidding. Gotta be by my caprine ball and chain, come what may. See you tomorrow.”

No sooner had she left, than the text notification chimed. It was the lab. Anti-freeze confirmed and certain pesticide residues. No need for remediation, as the concentrations were no longer sufficient to impact plant performance. However, that plot of land should be considered in transition for the next three years, and not fully organic.

Laurelyn was hugely relieved. She listened again to the message. Enormous quantities of anti-freeze over many years. That means the pit had been there a long time and was probably known by quite a few folks. Farmworkers, Jeffries' nephew, contractors applying any seed or fertilizer, and so forth. Would civil authorities responsible for dumping permits have known? Could there be complaints, or photographs, or other ways of discovery not involving firsthand knowledge? Knowing who to call on this track, she texted her friends Milton and Simon.

'Got another mystery, in case you hadn't heard. Call me.'

Chapter 4

Jack Splann was trying to be helpful. Really, he was. It was just so darned hard to remember those difficult times three years ago. Was the marriage troubled? Yes, but whose isn't? Were there financial and addiction issues? Not that he knew of. Did he have reason to believe his wife was unfaithful? Yes, he did. Did he resent it? Why sure, but they were working it out. Did he have a motive to kill her, no, because they were working it out. So, what did he think when she disappeared? That she had left him, according to the note she had written.

"And what note might that be?" Asked Mike. Although not a detective, he had spent enough time in the Los Angeles police force back in the day to know how to lay the basic groundwork and get some answers before Jack's attorney arrived.

"This note," the tidily dressed man volunteered, pulling an envelope from his pocket. Pulling a disposable glove from his pocket, Mike grasped the envelope then stepped to the door of the small interview room and called for an evidence bag.

"You carry that around with you a lot?" he resumed his interview.

"Why no, not at all. It hasn't been out of my desk since I got it in the mail. But I had a bad feeling about this interview, maybe an intuition I guess, so I stuck it in my pocket just in case."

"Very prescient of you. So, we should expect to find your prints, your wife's, and a postmaster's prints on the outside, your wife's and yours only on the letter itself?"

"I don't know, I guess so. Here, here is a photocopy I made so you can see for yourself what she said."

Mike took the letter, took note of the beautiful penmanship and the not so beautiful sentiment.

'Dearest Jack, I can't take life here in this town anymore. I was meant for better things. I am leaving you and will send an address and divorce papers when I get settled. Don't try to find me till then, you won't be able to. Yours, Tammy.'

"Pretty terse and right to the point. Did you ever hear from her again?"

"Never, and due to the nature of these questions, I assume she's dead. Look, this was a terrible time in my life that I am finally getting over. I will help all I can, but surely the police in whatever big city she was found will have a lot more folks than me to blame. Can I go now?"

"Not until..."

"Oh yes he can, unless you are charging him!" McGinty Faraday entered the room with his usual obnoxious flare. Although not Jack's usual attorney, he came highly recommended. Jack's regular guy was in the hospital after indulging in a political fundraiser packed full of people elbow to elbow. The pandemic that was decimating the nation had not passed him by, and he was currently in Grand Junction Regional Hospital on a ventilator. But this fellow, McGinty, was a friend and attorney of Milton Hornblend and had happened to be on this side of the Great Divide for a few days. Because of Laurelyn's text, Milton and Simon had a heads up and when Jack called Milton in a panic, wondering who had represented him with his own legal troubles the previous year, Milton had made the introductions.

McGinty had been in Marble taking depositions for a tort that involved a tourist who had tumbled off of the edge of the Crystal Mill. The plump lady from Nebraska had broken her arm and was suing everyone in sight, even though by all rights she should have broken her neck, given that she had squeezed under the safety rail. It only took McGinty two hours to wrap up his depositions and high tail it to Peony Creek.

"May I have the room for a minute with my client, Officer Horvath?"

With a shrug, Mike complied.

"What, if anything, have you said so far?"

Jack, being by nature a very precise individual, recounted the conversation word for word, leaving nothing out, and then handed McGinty his photocopy of the original letter sent by Tammy. He said, "OK, well that was plenty, more than plenty. We're going now. And yes, your wife's body has been found, but not in New York City or Boston, as you might have assumed, but right here in Peony Creek, so you can see why it is critical that you speak to no one on this matter and certainly not the police, unless I am right there with you. Let's go."

Back at Jack's office, a woeful tale unfolded. Of a naïve young man, thrilled to have snared a girl who was so plainly out of his league, only to find that he was the one who was snared. Ensnared. In a web of lies, infidelity, complaints, and insults that had made his marriage to that beautiful but deeply troubled woman a living hell. Nothing and no one was ever good enough for her, and she played the field with an abandon that would have made many a call girl proud. He spoke of her living far beyond their means, and yet not having any credit card debt.

"So you believe she had been engaged in numerous affairs with married men in this community? Got any names or proof to go with that belief?"

Jack said he suspected anyone with money or influence, but he thought a local banker, a former sheriff, and a district attorney in another county might all be counted. A high school math teacher had been fired for inappropriate behavior when she was a high school junior but he had left the area many years ago, three years before Tammy had disappeared.

"You're giving an avalanche of motive for you to be her killer. Now give me something that convinces me you're not."

"Seriously? But you're my legal representative, shouldn't my innocence be a given?"

"Should be, sure, but facts are what will win the day, not 'shoulds' or 'oughts'. So gimme, prove to me your innocence in order to make me the best advocate you could have on this side of Mars. But everything is important. So start with exactly where you were and where you had been when you received this letter. Did anyone see her after you last did? Was there any gossip or rumors that she was still in town, close by, or in another state or city?"

And so it went for the next four hours. By the end of that time, McGinty was convinced of the man's innocence but felt they had some big holes to plug for his legal defense. He photographed the letter with his phone and then scheduled another appointment for them later the next day.

The picture Jack painted would have given any mortal cause to strangle her. It was a very grim portrait of seemingly purposeful evil. It might be informative to talk to someone who had known her who was not a spouse or lover. Milton had worked in this part of the country for many years before settling down in Peony Creek. Perhaps he had known her? Milton being gay, she may have behaved differently around him, or he around her, and a different narrative might emerge. McGinty texted Milton.

The reply he received was negatory, but Milton knew someone who he thought had probably gone to high school with her. A follow-up text confirmed that Sally Handy, the owner of the herb shop on Main Street, had known her somewhat and was willing to speak with McGinty. Seeing that the time was nearly 7:00 PM and that he hadn't eaten dinner, he used his phone to text and email a request to meet with her the next day around noon. Soon a positive response was received, and McGinty then went for the largest steak he could order at a restaurant in Hotkiss known for its massive cuts of local beef, the Fork It Over.

McGinty didn't know it, but several other restaurant patrons and an employee were contemplating the same individual and her death. In a booth on the other side of the room, four men sat discussing their past, present, and future. News of the newly discovered body was now all over town, all over the county. The men at the table had all known Tammy Splann, three of them in the biblical sense.

Randy Miller, a local deputy district attorney had run unsuccessfully for district attorney three years ago, and fully intended to run again with the support of his loving and devoted wife, daughter of a local judge. His affair with Tammy had been brief but steamy. Eventually, he had broken it off and had been surprised how little this had upset her. A month later, he got a letter in the mail. Tammy claimed to be very upset after all, so much so that only $2,000 a month in hush money would keep her from venting her sorrows to his wife. Randy had paid that sum for nearly two years before Tammy's disappearance. He was very, very glad she was dead. This sentiment was shared by former county deputy sheriff, Alex O'Connor.

The big redhead had lusted after Tammy from the time she had been a flirtatious senior in high school that he had busted smoking a joint with a bunch of other 17-year olds underneath the athletic field bleachers. The other kids had run, but Tammy had slouched nonchalantly against one of the uprights. She snuffed out the spliff and crushed it into the ground. She approached the sheriff and informed him, with hand sliding down the side of his face, that if no word of this got to her parents, his rewards would be great. And they were. When his frumpy wife of 12 years was away visiting her parents the following week, they made love with such abandon that for him, it became just that. Love. After the week had ended and his wife had returned, Tammy informed him that all debts were paid and that they would not be seeing each other anymore. Alex was devastated.

During the time they had spent together, even though clandestine, he felt he'd never been more alive, never been appreciated and seen for the noble warrior he really was. Until Tammy. He had decided to divorce his wife, marry Tammy, and move away to a place where no one would care that he was 36 and she was 18. They only had to wait until the day she turned that legal age, just two months away. Therefore, her rejection via handwritten letter came as a blow to both his heart and his ego. He pursued her, hounded her, and generally made a fool of himself. Within six months his wife read the handwriting on the wall and left him. But before that day came, a second letter did. He was ecstatic, until he read the contents. Tammy wanted him to know that if he valued his marriage and his career, he would pay her the sum of $500 per month until she said otherwise. The marriage he could let go of, but his career, never. His anxiety and sense of betrayal eventually turned to rage and indignation. He paid the money, believing his career in law enforcement dead if convicted of statutory rape. But seldom a day went by that he didn't wish her dead, and had concocted elaborate plans to that end. When a chance for promotion and a pay raise in adjacent Mesa County presented itself, he went for it, hoping to distance himself from any scandal. And soon after that, the problem went away.

The First Bank of Peony Creek president sliced into his grilled sea bass filet. Clive Lancaster had pre-mature gray accents in his hair, a recent and appropriate addition to his stylish yet conservative wardrobe that had started by being born with a silver spoon in his mouth. Now there was silver on his head, a very distinguished-looking touch just at the temples. A Wharton School of Business MBA and a million bucks advanced to him by his parents 20 years ago, had enabled him to found a thriving, if small, local bank. Life had been a bowl of Royal Ann cherries, a Peony Creek staple crop, until his wife had developed MS six years ago. He loved her dearly, but her illness had robbed her of her most of her sight and all of her libido. Clive, a solicitous husband then as now, had grown sad and tired as her illness had progressed. A degree of loneliness set in, and no small measure of horniness. He had given everything he had in him for two decades to this community and to his marriage, and this is how God repaid him? He had met Tammy at a charity fund-raiser just one year before her disappearance and in a moment of weakness, had taken her to his bed. More than once. One slip-up, OK maybe four of those, threatened to destroy his modest empire.

He had been invited to a very, very uncomfortable dinner with the bank's accountant, Jack Splann, who made it clear that if he ever had a hint that Clive was still sleeping with his wife, he knew about things in the books that could become very awkward for Clive. And of great interest to the bank's board of directors. Clive, offended and outraged, yet knowing Jack to be the injured party, agreed in poor form, but agreed nevertheless. And he was as good as his word. Then a letter arrived. $3,000 per month required or his wife and her very conservative friends on the board would find out. Clive, swallowing a bite of the excellent entre, expressed himself to the table at large.

"Well, I for one am extremely glad the slut is gone. Good riddance to bad trash."

"Now Clive, that is unworthy of you, my friend. She didn't hold a gun to your head to sleep with her and last I heard it still takes two to do that tango."

Young and devastatingly handsome Reverend Lawrence Duckworth admonished his fellow diner. He too had almost fallen afoul of Tammy's wiles. God almighty, but she had been a beautiful woman, if extremely sinful. But a carefully laid trap had not been sprung. When Tammy had come to him for spiritual guidance concerning a mysterious addiction, he had been sensitive, kind, and practical. She had thanked him and left. She returned the next week wearing a slightly more alluring outfit, but nothing outrageous. On their third meeting, her low-cut blouse and push-up bra confirmed the sort of addiction she was suffering from. Weeping delicately, hand placed on his knee, she turned to embrace him. God, he was tempted, and he could sense that he could have her right there and then on the floor of his office. Tempted indeed, but once burned, twice wary.

He had gotten into trouble in his last parish with another attractive and married parishioner. He had been counseling her after a suicide attempt, and one thing just led to another. Of course, as God willed it, the husband found out and threatened to kill him. A grueling two years as a missionary in Africa helped to clean his slate and his résumé, and now for the past three years, his congregation was thriving, with no hint of scandal.

Therefore, Lawrence had carefully and tactfully disentangled himself and suggested that professional counseling might be better for her than pastoral counseling. A tiny indignant shriek, a slammed office door in his face. And that, he thought, was that. Wrong. Three days later he received a letter with a newspaper clipping concerning his past disgrace. The sum of $500 per month would keep her silent and prevent her from claiming he had tried to seduce her during their sessions. After a week of prayer, he decided on a course of action and he burned the letter. No hush money was ever rendered unto her, since she disappeared shortly afterward.

"We should pray for the repose of her troubled soul and hope that she finds mercy in God's forgiveness."

"Better his than mine," the cop intoned. "Sorry Reverend, but I wouldn't feel too terrible if she's roasting over Satan's eternal barbeque right this minute."

Whether Tammy was burning in hell or not, the ears of the waitress cleaning the adjacent booth certainly were. Deborah Tannenbaum Johnson, mother of Saffron Johnson, had recently relocated from San Diego. After six months of cosmetology schooling, she was beginning a new life.

Saffron did not yet know that her mother was now a Coloradan. They had not parted on the best of terms the previous year, although some small movements toward reconciliation had been made. And now the former stripper had purchased a little house for herself in Delta and had a part-time job in a beauty shop. She hoped her client base would continue to grow, but in the meantime, her faded but buxom bleached-blond look made her a shoo-in for a steak and seafood restaurant. And she was good at it. Her tips were actually better than her beautician income, at least for now. As she was strenuously wiping down the seat where a little kid had spilled a sticky drink, she could not help but overhear a portion of this conversation quite by accident. The rest she heard quite deliberately.

She didn't have to hear a name to know that they were discussing the deceased woman recently discovered on the farm property where Saffron worked. This went way beyond juicy gossip. This could be just the excuse she needed to contact Laurelyn, thereby reaching out to her daughter as well.

"Since you asked us all here, Reverend, I'm curious, did she ever try to get money out of you? Or did you dodge that bullet better than we did?"

"I hate to speak ill of the dead, but yes, she did, unsuccessfully." This admission in hushed tones.

"How much?" Randy Miller chimed in under his breath.

Lawrence looked extremely uncomfortable and finally whispered, "$500."

"Are you kidding me?" Miller burst out. "She was soaking me for $2,000!"

Clive looked green as he sat up straighter and groaned quietly, "$3,000." It was then that he realized that the waitress had been cleaning that table long enough to wear off the varnish.

"Excuse me miss, but could we get another round of drinks here?" When Lawrence looked about to object, Clive said qiuetly, "Somebody will drink yours, I'm sure, but I wanted to get that woman away from our table. I think she may have been eavesdropping."

Deborah was not entirely the bimbo she appeared to others. Before bringing the drinks, she quickly slipped her iPod and earbuds into the pocket of her apron. She brought the men their drinks and then made an elaborate show of plugging the earbuds in to do more table maintenance. This should give her a semblance of an alibi if anyone complained. Hear no evil. Once the men were engaged in conversation again, she took them back out and brought McGinty his check.

"There you go sir, and it was a pleasure to serve you. I have a silly question if you don't mind. One of the other girls told me to give you excellent service because you were a high falutin' lawyer, not that I wouldn't anyway," she giggled out for effect. McGinty smiled crookedly and said, "That is an allegation I can neither confirm nor deny."

Deborah laughed and pursued her line of questioning with, "So is it true that the person found up at that farm in Peony Creek died violently?" With that, McGinty frowned, and replied, "Now that I truly can neither confirm nor deny. Sorry." And with that, he put a $50 on the table, grabbed his coat, and departed. But he had answered her question. Only someone in the legal trade who knew the truth to be murder and not suicide or misadventure would have changed faces so quickly and thoroughly. This confirmation, that wasn't one, gave her the courage to phone the farm, and she determined herself to do that first thing in the morning. That resolved in her mind, she brought "the murder club," as she had started to think of them, their separate checks.

Chapter 5

Several things happened at 8:00 AM the following day. Tabby Ignacio sat at her desk and began her final report and certification of the cause of death. Laurelyn picked up her phone, answering a call from a number she did not recognize. And George rousted his son from his Bible reading and told him to get to work in the stable.

William Trojan was new to Christianity and as such, was very enthusiastic. William had been an enthusiastic drug user as a teen. Two years in the slammer had cured him of drug use and introduced him to something better. Jailhouse conversion had stuck for over a year now, as had sobriety, and his father was encouraged.

William worked hard in order to please the Lord. George wondered if William would continue to work just as hard to become a self-sufficient adult. But baby steps, baby steps. And William was pretty good with the horses. Tractor repair and maintenance was a bit more challenging, but Weston, bless his country heart, was nothing if not a patient instructor. In another few months, if William stayed upon the straight and narrow path, George was going to buy him a vehicle. But for now, William could only go where George or Weston went. It was definitely a test of his fortitude.

Today William would muck out the empty stalls all the way to the ground and then some. The corners had to be cleaned and the whole shebang disinfected with bleach in preparation for the season's horse-breaking business.

George never broke horses in the strictest sense, he merely coopted them to his purpose and leadership style. William found he could frame this in terms of God, Church, and Man. *Whatever works*, George thought, but kept it to himself.

Actually, the only time they fought one another was over religion. As any new disciple would, William wanted his father to take up the faith in exactly the same way he had. Only by doing that could he be sure that his father would enter the pearly gates. George assured his son that His Father's House had many mansions. William claimed that no man comes to the Father except through Jesus. George countered with the Eastern Orthodoxy that many of William's immediate ancestors had professed. William declaimed that they worshipped idols and therefore, like any and all Catholics, were going to hell. Eventually, George forbade discussion of religion under his roof.

William found this frustrating but realized that unless he wanted to be shipped back to his mother, who would promptly tell him to get a full-time job or leave, he had better comply. So he set his sights on another convert. Saffron. She reminded him of an angel, or what he thought an earth angel must surely be like. Her pale, ethereal appearance and delightful shyness were extremely appealing to him since she was just the opposite of the tweakers he'd hung around with in his misspent youth.

It had become plain to him that she had not yet accepted Jesus as her savior, but surely that was just lack of exposure to the Lord's teachings. Therefore, any time that George needed to go down to Bad Aspen,

William begged to go too. He did not always get his way, for one of the reasons that George was supporting his son was to be able to leave his ranch in good hands for a few hours now and then to court his sweetheart.

William didn't like Laurelyn. He didn't realize it was because he was jealous, and thought it was because she was an unrepentant sinner. A Jezebel really, in his opinion. When William had cornered her in the goat barn once, Bible in hand, and asked if she had heard the word of the Lord, he was horrified when she replied sarcastically, "Indeed I have, as Goat is my witness." Not entirely false, as her own parents were moderate Protestants, circumspect in their faith, but not the least bit evangelical. Laurelyn did not share their beliefs much, but didn't mind what others believed, as long as they did not espouse violence or hate. She found

William's arrogant assumptions a bit offensive, and although she too dismissed them as the follies of youth, she decided not to tolerate them.

After two unsatisfyingly similar encounters, William desisted, consigning her to hell where she probably belonged. He just wished his dad wasn't so smitten with this whore of Babylon. He would make it his holy and familial duty to redirect his father's interests in the direction of a nice Christian lady who wanted marriage, like the one he'd gone out with before. That prospect was no longer available, both since Sally had chosen the primrose path and further, was currently dating Weston. But surely in a town that boasted having held the Guinness Book of World records for the most churches per capita in the United States, there were righteous women. Once he had his own car and could go to

Sunday services he would see if he could make friends with some nice Christian girls and their widowed mothers.

It was these comforting certainties that kept his demons at bay. That religious zeal was his new drug would never have occurred to him, although it had occurred to his mother, and was becoming apparent to George. But George felt that unless he was radicalized into a suicide bomber, it was a better addiction than most. Internet-based radicalization at George's abode was unlikely, since his Internet satellite reception was spotty, as was cell phone service. Face-to-face recruitment was another thing altogether, which was why William had to prove himself for a while longer before receiving the vehicle which might transport him to religious gatherings led by persons eager to take advantage of his youth and naivete. But until then, his labors at the Trojan Ranch would have to do as penance for past sins.

"William, you good here?" George hollered through the stable entrance.

"Doin' good, just two more stalls to go!"

"Great, cause I'm going to pop down to Laurelyn's for a while. I wanna make sure she's holding up OK with that body discovery and all."

"Oh man, can I go too? I will finish up here fast, and maybe do the last stall later? I really want to talk to Saffron."

"About what?"

"Oh you know, just stuff."

"Religious stuff?"

"Yeah, maybe that and some other things. I want to see how she's doing, too. Laurelyn's not the only one that lives there, ya know," he added with a ghost of his early teenage snottiness. Instead of enraging George with its disrespectfulness as in days of yore, it made him smile a little. It seemed like a semblance of normalcy, a dingy and shallow but familiar port in a spiritual storm.

"Sorry son, not this time. There's two horses arriving tomorrow morning and our days are gonna get pretty full and I need all those stalls done this morning, since we will be oiling, cleaning, and inspecting tack this afternoon. Plus, I need you to keep an eye on Shamrock, since she's due any minute. I'll tell Saffron you said hello, though."

"Yeah, whatever," he groused.

That second comeback did make George a little tense. Maybe this experiment wasn't going to work after all, and the first cracks were showing. Thrusting that aside, he climbed into his truck and headed down the hill. William, genuinely regretting his sass, sat on a cold bale of straw and bowed his head in prayer.

McGinty and Jack were sharing waffles in Jack's kitchen at about that same time. McGinty was going to have to head back to Denver for a couple of days shortly, and wanted a better accounting from Jack of just what he knew of Tammy's peccadillos, how, when, and with whom. Why would be nice too.

"When she said yes to my proposal, I thought I was the luckiest man in the world. I mean, I'm not captain of the football team stuff, or movie star leading man either.

But before the end of our first year of marriage, I knew something was off. I could just tell."

"Tell how?" McGinty pressed.

"Odd little things like putting on perfume to go to the grocery store, or wearing a low cut blouse to a charity board meeting where everyone there was a conservative and religious local woman over the age of 50. Things that would be innocent enough under certain circumstances but were very odd in others. Then I started noticing odd bits of jewelry that I didn't recognize. She would tell me that I was being silly and that she'd always had those. At first, I took her word for it, but after a certain point, I knew someone was giving her things. There'd been talk of problems with her high school math teacher, so I assumed it must be him. But like I said, I was able to confirm that he left this area years ago and was happily ensconced in Florida, making a fortune as a stockbroker. It hurt so much to realize that there was someone new that I refused to see it as a pattern or pathology. Now in hindsight, I know better."

Over the course of the next hour, and much prying, McGinty was able to come up with no less than nine people, Jack included, that might want Tammy deceased. And that wasn't even counting the right Reverend Duckworth, whom neither Jack nor McGinty knew about yet. It was unlikely that the wives of the adulterous men had the strength to manually strangle the tall and athletic Tammy, so that reduced the suspect pool a little. Unless of course, a wife had hired someone to do the deed. This gave McGinty plenty to think about as he drove over to the Merry Clary to interview Sally Handy. McGinty did not know that the nosy waitress from last night's dining was on the phone to Laurelyn at that same moment.

Chapter 6

Deborah's hands were glistening, shaking, and just about useless. Nevertheless, she put her fears of rejection aside and managed to dial the number. After four rings she nearly hung up, but on the fifth ring, the pleasantly confidant and moderately puzzled voice of Laurelyn answered. Laurelyn had glanced at the phone, and seeing it was an unknown number she had assumed it was a robocall and nearly didn't answer. But then seeing that it was a San Diego pre-fix, she took a chance it was a legitimate call. A chittery, broken voice on the other end identified itself as, "Hi, I mean, hello, I'm Edith's, er, Saffron's mother, Deborah. I was hoping to be able to speak to her and you about something important, at least, well, I think it's important, but not sure yet, so, ah dammit, I might have news about the murder."

A second, then two, and finally Laurelyn said, "Bambi?"

"That was my stripper name, I'm Deborah now, remember?"

"Oh, that's right. How have you been? Are you OK?"

"I'm great and I live in Delta now. Saffron doesn't know cuz I haven't had the guts to tell her till now. But I heard something last night at the Fork It Over that I think you should know."

Right at that moment, Saffron entered the greenhouse to get further instructions for the day. A furrowed brow and Laurelyn's double glance made her hesitate. She hoped it wasn't George canceling his visit for this morning. She really wanted to get his input on how her dedicated garden was coming along. The gazebo was done, and she wanted to show it off.

Placing the phone against her thigh to muffle the sound, Laurelyn said to Saffron, "This call is actually for you."

"George?"

"No... your mother."

Seeing her face start to pale, Laurelyn awkwardly but firmly thrust the phone at her. She thought if she gave Saffron a chance to bolt, she would, and any potential reconciliation would be delayed a very long time. Besides, she definitely didn't want to get between these two. She had become a surrogate mother for the girl and was very protective of her, but family is family and if there was a glimmer of hope, she had to facilitate it as best she could. It was the only righteous course of action.

Laurelyn didn't get to listen in on the Saffron side of things, because she could see George striding down the path from the parking lot. She wanted to head him off, figuring it was one thing if Saffron's mentor kibbitzed, another thing if there was a growing audience. She grabbed George and with a quick but passionate kiss, escorted him to Saffron's Garden. She apprised him of the situation. After walking around the gazebo and discussing plans for dinner in town soon, he asked, "So that's Bambi Johnson on the phone? She's not planning on causing trouble for Saffy is she?"

"It doesn't seem that way yet. It appears that she overheard some very juicy gossip about the murder last night and is using it as an excuse to reach out. She lives in Delta now, of all the darn things."

"You're kidding me! I sure hope this doesn't go south and fast. How do you feel about that?"

Sigh. "Not sure how to feel just yet. It may take a bit. Speaking of adult children, how's William these days?"

Sigh. "A little snippy this morning. He wanted to skive off his farm chores and come down here with me. I said no and he gave me the first lip since he's lived with me. I'm sure he's flagellating himself right now for displeasing God and not honoring his father."

"I hope he is! Although I have wondered if all this religious fervor is really that healthy for him."

"You and me both, but what are ya gonna do? I made a six-month commitment, and he has less than three months to go before I can pull the plug. But he has made huge strides, compared to who he used to be, so I'm crossing my fingers concerning the prodigal son. Now, this gazebo needs a little gingerbread in some of these corners. I think I will buy a few scrollwork pieces and install them myself as a garden opening present for Saffron," he adroitly changed the subject.

"That would be very sweet of you and I bet she would be thrilled," Laurelyn replied, taking the hint to abandon the topic for now. They both noted the arrival of Beth who entered the chiltibarn to begin watering, labeling and lightly fertilizing some emerging young peppers. Soon Saffron appeared. She handed the phone to Laurelyn with shaking hands and threw her arms around George's waist.

"Whoa there girl, easy now. What's all this?"

Not finding the words, which was often the case when her autism kicked in during stress, she gave him a final squeeze and headed to the goat barn.

"Probably going to hug Belle and sit with her and the baby goats for a while to get centered. I'll give her some time and then try to get her to relay the information that Deborah imparted."

"You want me to do it?"

"Very kind, I but think this may need to be a maternal mash-up. No dudes allowed just yet. But you will be the first I tell afterward."

And that would have to do the guy for now. Weston arrived and together he and George rearranged some hay bales that looked like they were going to need to be fed out sooner than hoped, then covered the rest well with tarps, since it was likely to snow tomorrow.

At the Merry Clary, Sally had just gotten the store opened and the new teas stocked when McGinty entered the door. She had been expecting him later and hoped she didn't have any customers for a few minutes.

After introductions, McGinty took command of one of the chairs in the small reading nook and in best courtroom voice, began his interrogation with, "So how well did you know the deceased?"

Sally was taken aback. She stammered a little about not being best friends or anything with Tammy. McGinty pointedly looked at this watch, as if she was holding him up, and there he was trying to do her a favor. That made her mad. Sally was seldom angry, but when she was, she was quick and rude to let someone know it.

“Look, Mr. Faraday, I don’t know where you need to be or honestly, just who you think you are, but if you want my help any further, then cool your heels, buy something, or leave.”

To her surprise, McGinty smiled. He knew he was a jerk, and that his social skills sucked. But his courtroom skills were superb. He knew he would now get a really honest answer to the next question.

“Do you have any reason to think that any particular person in this community might have a reason to kill her?”

Sally hesitated. Her immediate response would have been, “Who didn’t?” But that was not the fullest answer possible. She centered herself and said, honestly, “I always got along with her fine, but there were many who didn’t. Tammy, um, well, I guess you could say she marched to her own drum, to a beat that most other people disliked. Hated even, sometimes. She was never a virtuous woman, but she did have a kind streak every so often, especially for animals, children, and various underdogs. Some people thought she was slutty, and maybe she was, I dunno. Given her looks, just about every married woman in this valley had reason to want her gone, one way or another." With this statement, she paused, hoping she had given him what he came for and would go away.

"Names?" was the only thanks she got.

"Sorry Mr. Faraday, I'm not going there. You asked me to talk about my impressions of Tammy. We were friends, in a way, and whatever others may have thought of her is none of my business. I plan to keep it that way as much as I can. My impression of Tammy was that she was a lost soul, and that's all I have to say on the matter. Now if you will excuse me..."

McGinty left, without a thank you and with no purchase, as was his usual way. She had given him a tidbit that his client hadn't really mentioned. The jealous female of the species might indeed be the murderer. Either a large and athletic woman or a hired assassin? More information on that angle might make for an interesting argument in favor of his client's innocence. Or at least throw plenty of doubt on his guilt. With those thoughts, McGinty climbed into his Suburban and headed for Denver. He took the scenic route back towards Marble and passed the turnoff to George's ranch on the way.

At said ranch, William had prayed himself out of his funk and was working extra hard in the stables. His penance was so thorough that a human could safely eat off the floor, let alone a horse. George, returning from town after Laurelyn's, grocery bags in hand, entered the stable with trepidation. He nearly had to pick himself off the carefully raked sawdust-laden corridor. The smell of bleach permeated the air. In the four hours he had been gone, William had done the work of two men. George smiled with relief and fatherly pride. Other than fresh shavings in the stalls after the bleach had aired out, nothing further needed to be done there before the first two horses arrived tomorrow morning.

An off-the-track Thoroughbred three-year-old being started in dressage, and a Morgan Arab being schooled for endurance were scheduled. The owners of the two were fussy, rich Colorado transplants from Florida, and only cleanliness at this level would do for them. Since they were shelling out $1,400 per month per horse for board and training, they were entitled to having their expectations met. George would take only two more horses for schooling for the rest of the year, so as to give the animals the attention they deserved. William would be a huge help with that.

William, being of slighter build than George, was tasked with working with the thoroughbred. Although the horse was over 16 hands, his youth meant that his bones were not solidly formed, and the less weight on his back, the better. The other horse was a four-year-old in need of a very firm hand. George's specialty.

As George carried the provisions up to the house, he thought that a reward might be in order for the lad.

“William, good job on the stable, son. Say, I have to go back down to Bad Aspen late this afternoon. Wanna go too?”

“Well duh, I mean, of course, thanks, Dad.”

He didn’t really have to go, of course, but was dying of curiosity about the sudden appearance of Saffron’s biological mother. So, he could kill two birds that way…reward his son for valiant effort and satisfy his curiosity. And take his beloved two lovely bottles of Pinot Grigio he had just acquired.

Said beloved was sitting in the goat barn with Saffron. Without speaking, they played with the baby goats, while watching the fourth doe preparing for labor. Laurelyn felt that some form of verbal processing of this familial development would be a healthy thing. But with a traumatized young woman with autism, maybe that wasn't so? She wondered if she should call licensed social worker Nancy Clifton for advice, Nancy having prior experience with both Saffron and Bambi/Deborah. If she couldn't coax any words from Saffron in the next two hours, she thought she might make that call. A grunt, groan, grunt, SHRIEK, interrupted her internal discourse. Something else internal was about to become external, and from the sound of it, might be complicated.

Two hours later, the doe had made little progress. It was almost certainly going to be another breech birth, unless this baby was huge. When George and William showed up, she was incredibly relieved. She had needed to use the toilet for over three hours, and now was her chance.

She stood next to George, making her apologies and promising to be right back.

"Bring two wine glasses, would ya?"

Only then did she notice that George had a bottle of cold wonderfulness in his hand, and William had a Bible in his. She frowned at that but said nothing. Saffron, standing more or less next to William, had noticed it right away, and was even more offended by it than normal. Understandable, under the circumstances. When William opened the Bible to begin a blessing from some portion of the Old Testament, she kicked him in the shin with all the might of her little form. Her half-steel-shanked work boot made the impression that the force of the kick lacked. He howled and exploded, "What the hell?" then clapped his hand over his mouth like a 6th-grade schoolgirl.

Saffron started to apologize when a blousy figure treading its way down the path caught her eye. A simultaneous shriek from the doe and the young woman made a weird stereo effect that would have been perfect for terrifying Halloween trick or treaters.

Saffron bolted from the barn, not out the door that people used, but out the open end that led to the pasture. She vaulted the fencing with an Olympic grace and proficiency that caused all watchers to stare with incredulity. All watchers in the barn, that is. The approaching figure did not witness the girl's flight, as was the intention. In tears, Saffron made her way to the house unnoticed by her mother.

Laurelyn sighed, her bladder bursting. Putting even more burden on George, she enlisted his help in unwanted visitor wrangling before she wet her pants, and then sprinted in bow-legged fashion up to the house herself. Deborah stared at her as she flew past. An attempted arm grab failed, but spying George, she continued on with her hejira of supposed reunion with her only child.

Sitting on the potty, hearing Saffron's sobs, she thought to herself, *Definitely calling Nancy*!

Chapter 7

Sandy Horvath, ne Emmons, sat at her computer ordering supplies for the coming tourist season. Outside her window on the second floor, she could see her husband Mike continuing the vine binding job she had abandoned yesterday. He was such a good man, and even though they didn't see eye to eye on everything in life, well criminy, who did? But now that she was in possession of information concerning a murder that she really shouldn't have, was she betraying their relationship? Was she putting her friendships ahead of her marriage?

Her marriage and her friendship with Laurelyn had been almost simultaneous in her life. To be fair, Mike had preceded Lor by two years. But there are ties that bind that have nothing to do with romance, and her friendship with Laurelyn, and through Laurelyn, Bernie Gonsalves, was a chick thing for sure. And that was not to diminish its strength. It was just a horse of a different color from that of a spousal relationship. Ironically, if it hadn't been for Mike, she would never have met Laurelyn or Bernie.

When she had met Mike in criminology class and they had begun to date, it had been passionate and surprisingly meaningful. When Mike asked her to marry him after dating her eight months, she had said yes without hesitation, and never looked back. When they made that mutual decision about only one cop in the family and she had switched majors, that switch had led her to Laurelyn. When Laurelyn had introduced Sandy to best friend and former fellow zookeeper Bernie Gonsalves, the two of them had bonded over law enforcement stuff.

Sandy remembered those days in the greenhouses on campus. She was Laurelyn's mentor, and the three women had hiked and feasted and partied with gusto anytime Bernie was in town. So as Sandy watched with dismay the slow, skillful seduction of Laurelyn by her main professor, and with growing horror their marriage and Stewart's fast and inevitable betrayal, she did the only thing she felt she could do. She ratted her out to Bernie.

In usual Bernie fashion, retribution on the feckless Stewart was swift and ruthless. The two friends then moved Lor to Peony Creek for a fresh start. And the start itself had been brilliant. It was just the whimsy of a capricious universe that Laurelyn had been forced to cope with three murders in less than one year's time.

Mike had supported everyone's efforts all the way. Remarkable for a semi-macho man. Semi, because he had embraced none of the toxic masculinity, sense of entitlement, or resentment, so prominent among law enforcement males. So noticeable was this nobility among friends and associates, that upon his retirement party (after five blood-curdling years in the Ramparts division of LA) his co-workers had given him a British Bobby's hat and whistle, and referred to him as Constable from that night. And he had laughed his ass off but did not demur. He was rather flattered, in fact.

They had moved to Peony Creek just so that he could actually be Andy of Mayberry. And that had worked for a while, but these last few months and the body count had been hard on him personally and hard on them as a couple. They sometimes wondered if they had bitten off more than they could chew. But they persevered. The community helped, their friendships helped, and their sense of purpose helped.

Sandy knew all of this and knew her husband's extraordinary strength and flexibility. But at what point would he object? Feel isolated, betrayed, or excluded from Sandy's inner circle? She had no one to turn to with those questions. Laurelyn was too wounded by widowhood followed by intense spousal betrayal to be a font of wisdom, and Bernie, was, well, Bernie. Gorgeous, irreverent, violent Bernie chewed up men and spat them out regularly. Sandy was the one shining beacon among the three for what could be, should be, one day with the right partner, down the road. Sighing, she made the difficult decision to keep all news about the murdered woman gleaned by the Tres Amigas to herself, at least for now. She hoped that one day he would understand.

Determined to investigate further, she texted Simon. She knew the young public defender had made a surprising ally in the new medical examiner. Since they were both ethnic minorities and had fought unusually hard for their positions in this mostly white community, Tabby had grown to both like and respect the gay, Asian civil servant. She was never one for bending the rules much, but when a little info on a case could not possibly harm either the defense or the prosecution, well, sometimes little nuggets came Simon's way. Sandy was hoping that the actual cause of death, which would be made public soon anyway, might qualify as one of those nuggets. Simon texted back he would slyly address her inquiry.

An hour later, she received a reply. The cause of death was manual strangulation by a medium to large-sized person with strong hands. Sandy's thinking ran parallel to McGinty's at that point. Virtually impossible for a woman to have done it, unless in a secondary fashion, with an accomplice or a hired assassin. Assassins in this area were about as common as the Mafia, so maybe an accomplice personally known to the victim? That said, the most parsimonious explanation was that the killer was known to the victim and was a strong male individual at least six feet in height and probably 50 years or younger. Another text and image from Simon came in. Simon had been discussing the case with Milton, and Milton, hearing that there was a note involved, asked McGinty to forward a copy, which he did. Sandy's eyes lit up. Now surely that was one piping hot clue!

As she was reaching for the phone, she got a text from the very person she had been planning to contact next. Laurelyn had typed, '*Bambi here now, please come.*' A second text followed. Knowing that with inventory done, she would have to launch into taxes, due in just a few days, Sandy was thrilled at an excuse not to venture there. She grabbed a coat and raced out to tell Mike she was needed at Bad Aspen and would see him at dinner. He grabbed her and kissed her hard, then warned her not to get herself into trouble.

"Oh, and can you see, sweetheart, if you can get on the horn to Nancy Clifton and ask her to head up to Peony Creek? Her professional services are needed, and

Laurelyn can't get ahold of her." She knew that a request from law enforcement got higher priority than one from the general public. And it gave Nancy a certifiable excuse to drop other engagements and pursue the request, should her actions ever be questioned in court.

On her way down to the farm, she mused that at 5'11", husband Jack could be the killer. But even though he was a healthy and vigorous man, he was not terribly fit, spending most of his days in an office. But if the body had been there three years or so, what was he like back in the day? Had he been more athletic? Sally Handy would know.

As she drove, she dialed the Merry Clary.

"Bright greetings to all, and how may I assist you in your journey?" Sandy winced at the New Age lingo but persevered.

"Hi Sally, this is Sandy Horvath. My path today has led me to ponder how much you know about Jack Splann. Was he ever an athlete, back in the day?"

"Chris Chringle on a cracker, not you too!" an exclamation that Sandy found indecipherable. "Sorry, but Jack's attorney was just here asking me questions about Tammy. Anyway, Jack was always bookish, head of the debate team and high school comptroller and such. Never played any team sports that I'm aware of, then or now. Why do you ask?"

"Oh nothing, Laurelyn thought he might be and was going to send him a sporty condolences card. I will advise her to reconsider." Sporty condolences card, seriously? That was the stupidest lie-on-the-fly she had ever invented. But Sally seemed content with that explanation.

"Yes, that might be kinda tacky. But I have a very nice selection of cards right here if she decides on something more appropriate."

"Thanks, I will let her know."

So, never an athlete or bodybuilder or anything like that. Sure, rage and adrenaline could give remarkable strength to one's hands, but not to the length of one's hands. From Simon's details of the ME's findings, the extent of the bruising indicated very large hands. She found herself exonerating Jack more with each passing mile.

When she arrived at the farm, she headed down to the goat barn. She had a feeling that would be where the action was, and she wasn't wrong. George and William were all riding herd on Deborah, to keep her from searching the property for her daughter. Sandy sized up their strategy, which wasn't working particularly well. She greeted the group with overall cordiality. She also saw that a doe was in difficult labor. Caught betwixt the need to keep Deborah away from Saffron, and to get her away from the goats, she was flummoxed. There were way too many people in the barn just now. Then inspiration struck. With a woman's insight, she saw the need to substitute one motherly impulse with another. She pretended to notice something wrong with one of Belle's kids.

"Oh my gosh, that little girl is so cold! She needs to be warmed up. Deborah, would you mind holding this baby for a little while so I can go into the milking room and find a coat for her?" Although coats were indeed available for newborns who seemed chilled, this baby was actually fine, but Deborah could not possibly know that. Without waiting for agreement, Sandy marched over and picked up the little girl and snapped a baby goat coat onto her. Belle, trusting as always, took the kidnapping in stride.

Sandy put the baby in Deborah's arms after showing her how to support the little one's legs. It worked like a charm. Stiff and uncertain at first, Deborah began to relax and all thought of searching the farm for Saffron evaporated. They put coats on all three babies and then Sandy, leading Belle, took all six of them all into the adjacent paddock with Thor and Titan. The two enormous wethers started to approach to see if there were any treats for themselves, but facing Belle's lowered head in warning, and getting no encouragement from Sandy, they left their corner of the barn and went back out into the sun. Laurelyn kept a couple of rounds cut from enormous old cottonwood trees for people seating and goat jumping in the boy's pen, and the ladies sat on these with their bundles, while Belle contentedly looked on.

George recognized this clever ploy and gave Sandy a surreptitious thumbs up. But this excess of womanly sentiment made the men uncomfortable, who promptly decided to find other places to be. William made as if to go up to the house to be with Saffron. George stopped him with a scowl and shake of the head. He also held out his hand for the Bible.

"Time to put that away for a while, son. God will understand," he improvised. William clutched it to his chest like a drowning man with a lifesaver. George sighed.

"You have two choices, you put that in the truck and we stay here, or you hang onto it and we leave. You choose."

William was torn to the roots of his soul. Surely his own father was not asking him to make a choice between himself and God! He looked up at the house and then at his father's determined, yet oddly kind expression. He worried that his father was indeed going straight to hell and guaranteeing that Saffron did too.

George, seeing the conflict, felt compassion and had a sudden fatherly insight. This conflict was really not necessary.

"William, don't you have that thing memorized by now? What sort of disciple of Jesus doesn't know the most important scriptures by heart? You don't need a book as a prop. Your faith should be up here," he tapped his forehead, "and in here," he tapped his heart.

William was thunderstruck, humiliated, and then inspired in the space of five blinks. He handed his dad the Bible with reverence.

"Good on ya, son, you have what it takes in your heart. Now, I see that Nancy Clifton, that social worker I mentioned, is here. She might want to speak with Deborah or Saffron without us around. How about if we go and replace that spark plug on the tractor? You replace it, and I will supervise. Then, when the time is right and everybody is sorted out, I bet Saffron would love to talk to you, just you, without the Good Book coming between you." George, fearful that he might have overplayed that hand, watched his son's face.

Doubt, then clarity displayed, William shook his head in earnest agreement, and off they went, George's arm around his son's shoulders for the first time ever.

Deborah noticed none of this exchange. Her entire focus was the baby goat, and she sang to her and rocked her like a human infant. All of which the kid enjoyed immensely. Sandy congratulated herself on her own erstwhile motherly instincts. *Not bad,* she thought, *for someone who never had children and is never going to.* Then she quietly stood up beside Deborah, beaming approval.

Meanwhile, Nancy had been apprised of the situation by Laurelyn. She was torn between counseling Deborah on the time these sorts of reunions can take and checking with Saffron. Her decision was made for her when the Riveras pulled into the parking lot.

Beth had texted them. Their daughter, Tiffany, a Down syndrome child that Saffron was very close to, was the closest thing to a best friend that Saffron had. Actually, because of their age difference, Saffron was subconsciously practicing the parenting skills that she was learning from Laurelyn. Patience, thoughtful guidance, appropriate spontaneity, and acceptance were something Saffron had been denied as a child. Tiffany was blessed with that nurture in abundance, but her parents always appreciated a trusted soul to give them a break from her needs now and then. Saffron was their babysitter of choice, and they were more than happy to assist any way they could.

Nancy, something of a social services ninja, sized all of this up and headed for Deborah. Along with Sandy, she cooed and approved of Deborah's way with the baby goat.

For her part, Deborah was on the threshold of a healing within herself. An opportunity to hold a non-judgmental, fragile little life was allowing her to stop beating herself up about the lousy parent she had been. It would take years still, but this moment, this unexpected moment, was the one she would remember as the true start of that process. She had needed to beat herself up, and she had, and rightly so. Now it was time to move on. Nancy sensed this and wanted to see how it played out as therapy. Sandy slipped away, knowing she was needed more elsewhere.

"Where's my baby girl? My little Pookie? You can't keep her from me you know!!"

"Ms. Johnson, no one is hiding Saffron from you. She turns 22 next month and you know how your guardianship maneuver last year went down in flames. So, take a breath, and tell me why you are here again, making demands you have no right to make."

Deborah deflated. She was used to hiding behind charm and/or the bluster of her abusive husband. Six months wasn't nearly enough for her to re-invent herself, but she was trying.

Nancy sat on the vacated tree round and pulled Deborah back down beside her. She then snagged a baby goat for her own lap and they began to process. Deborah recounted her journey from battered arm candy to beautician and waitress. It had not been an easy path. She had a high school degree and no resume, other than as a stripper. Nancy found herself reluctantly admiring the woman. There was strength and tenacity under that flouncy fluff, and Nancy made a private resolution to herself to help Deborah succeed as much as she could. To that end, she made a hair appointment with her for the following week.

Feeling as though she had a new friend and potential ally in Nancy, she recounted in excruciating detail the overheard conversation. Nancy's eyes grew wider by the second.

"Do the cops know any of this?"

"No, I wanted to tell Saffron and Laurelyn first, but you are so easy to talk to, and Laurelyn seems stand-offish, so I guess I just unloaded on you. Sorry, but now someone besides me knows what was said. Oh, and here are the separate receipts they signed for their meals. Separate tabs, from grown men. What a bunch of pussies. Oh, sorry, that's rude."

Nancy smiled and took the receipts from her. Putting the baby goat down, she took the receipts into the milking room and snapped individual photos of each one with her phone. Returning them to Deborah, she said, "Better get those back into the till before the restaurant's bookkeeper misses them. Let's keep this just between us for now. So here's a piece of advice and offer of help, all for free. Don't try to see Saffron today. Give her some headspace to adjust to your new proximity. Go on home and call tomorrow to make an appointment to see her. No more drop-ins. In the meantime, I'll have a chat with her to help get her on board with a renewed understanding of who you are now. I think she will be receptive. But after that, it will be up to you guys to navigate a new relationship. Slowly. It seems from what I heard that there was a hint of that at your last meeting. Build on that, sure, but be patient. Saffron is an amazing young woman, but she is still a person coping with autism. So baby steps, baby steps," she advised, not knowing she was channeling George's advice to himself concerning William.

"Oh, and one last thing, be willing to take no for an answer. She is a fully functional adult and can decide who she does and does not want in her life. Be ready to back off if need be and start again at some point in the future perhaps."

Wiping a tear from her eye, Deborah kissed the baby goat and put her down. A deep breath, a brushed off skirt, and she headed for her battered sedan.

Chapter 8

George and William watched Deborah return to her car, and saw Nancy head up to the house. Shortly after that, Saffron emerged. She straightened her thin shoulders, and with a determined tread, headed toward George and William.

"Nancy said I should apologize to you, William, so I am. Sorry. But we at the farm all agree that you need to keep your religion to yourself if you want to be welcomed here. Sorry." And with that, she proceeded to the chicken house to collect eggs with Tiffany, who stood there waiting for her with her usual lopsided, joyous expression.

George lifted hands in a 'whatever' shrug that said it all. The tiny smirk on George's face was instantly erased and replaced with a serious expression when his son glanced at him for backup. "I can't support you on this William, because it's their place and their rules. Can't you just enjoy the people and animals here without having to convert them?"

"But Dad, I love all these people! Especially you, and Beth, and Saffron. I can't bear the thought that you are going to hell! OK, I might be fine with Laurelyn, but not you guys! It's not right." And in an astonishing display of emotion following that righteous outburst, he collapsed into tears and raced for George's truck up in the parking lot. But not before yanking the Bible out of George's hand. In his wake, George and Saffron stared at one another across the yard, stunned and speechless. George walked over to her.

"I'll go and talk to him. I may need to take him home, however. If I do, could you please tell Laurelyn that I shall return?"

"Of course, be happy to. The poor little thing, I shouldn't have kicked him."

George smiled at Saffron calling anyone 'a poor little thing.' He gave her a quick hug and went to join his son. Meanwhile, Nancy, Laurelyn, and Sandy were enjoying a well-deserved glass of wine by the kitchen window when William went streaking past the house.

"I seem to bring joy wherever I go."

"Now Nancy, behave, this situation was fraught with anxiety and hurt feelings long before you entered the scene," Laurelyn assured her.

"I dunno," Sandy appended. "She does seem to have a certain wicked way about her. I believe her soul may be in jeopardy."

"Enough of that, heathen wench," Nancy retorted. "Seriously though, I have a niece that is just like William. She is really sincere in her beliefs and her concern for her fellow man but was way over the top with her rhetoric at family gatherings. I have to say, however, that in the last year she has mellowed out some and seems more comfortable and tolerant of conflicting beliefs held by others. She appears to be maturing into her new faith, and I'm guessing that with time and patience, William will do so as well."

"I hope so," Laurelyn sighed. "He seems to hate me pretty thoroughly. George has not let that come between us, but blood is thicker than romance and who knows how long he can hold out."

"That man? Oh, you need have no fears there. He has a will of iron, and his opinions on most things are equally durable. William shall not prevail against those battlements."

"I hope you're right, Nancy. So, you wanted to see the burial site, did you? A macabre part of your psyche I never knew existed. Oh good, here's Beth. She won't want to go, so I can get her to supervise Saffron for a while, maybe get her to take Tiffany on a tour of the new gazebo, the one Beth paid for. That seems appropriate."

A shriek from the barn interrupted their reverie. "Oh My God, I so have to get down to the barn again. Come on ladies, lend a hand."

The three women, glasses in hand, scuttled down the barn. But for the most part they weren't needed, as the three-year-old doe, a second freshener, had finally birthed on her own. Two enormous baby boys, the biggest so far, lay in the straw. Though both would be of equal height one day, one was fat and one was thin. While proceeding with toweling and navel dipping, Laurelyn announced, "I now pronounce thee Laurel and Hardy. Weston?" seeing he had just arrived. "Oat broth!!" The tired mother was already licking the boys enthusiastically. They were gorgeous, and no sign of any leg issues. "Stay here team, I will be right back and then we will satisfy Nancy's morbid curiosity."

She quickly sprinted to the herb and vegetable greenhouse. Beth was separating the heap of planting baskets that had just arrived via UPS. Her intent was to begin growing hanging color arrangements in abundance this year, for sale and for show during the farm's two open field days. The Spring Fling day would happen in just 10 weeks. Barely time to get those baskets going, but with Saffron's help, Laurelyn had no doubt that the march of time was no match for those horticultural warriors.

Saffron, Beth, and the Riveras were all happy to skive off the mordant march to the poison pit, so they all went to examine the gazebo. Tiffany was always happy with whatever. The adult Riveras tagged along, happy to provide emotional support. They did 'oohing' and 'aaahing' really, really well.

The two sleuths and one gawker made their getaway and headed out to the rectangle of yellow crime scene tape.

Ducking under the tape, and standing around the edge of the pit, gazing in at the green slime on the bottom, Laurelyn noticed something. It looked like a slip of paper with writing on it. As she edged to the edge and squinted at the object, the ledge gave way. She unceremoniously slid on her butt to the fortunately shallow bottom. Her calves landed on either side of what turned out to be a sharpened piece of rebar. It was plainly intended to cause serious injury or worse. She stared at the stake, and the horror of what might have happened was eventually replaced with gratitude that it hadn't.

Carefully righting herself an inch at a time, in case there were more hidden pikes, she plucked the note from the stake. She gasped and handed the note to Sandy. Sandy cautiously took it by one grimy corner.

Pulling a Kleenex out of her pocket she cautiously held the note in the sunlight and took a picture with her phone. Gently folding the note in half, she placed it in a shirt pocket. Only then did she dial her husband.

"Mike, honey, could you get to Bad Aspen right away? Someone just tried to kill Laurelyn!"

Chapter 9

While Laurelyn was climbing out of a death trap, Reverend Duckworth was seeking inspiration. He had a sermon to write and his thoughts were all over the place. They did, however, keep coming back to one topic, and not a sacred one. Tammy Splann. He shook his head, trying to concentrate. Scapegoats kept coming to mind. Is that because Tammy's body had been found on a goat farm? Well, it was nearly Easter after all. A traditional sermon on scapegoats was generally delivered about this time of year. He opened his Bible to Leviticus and read...

'And Aaron shall lay both his hands upon the head of the live goat and confess over him all the iniquities of the

children of Israel, and all their transgressions in all their sins, putting them upon the head of the goat, and shall send him away by the hand of a fit man into the wilderness. And the goat shall bear upon him all their iniquities unto a land not inhabited: and he shall let go the goat in the wilderness.'

Feeling guided to craft a homily recounting this mystical yet somehow seemingly appropriate cultural metaphor for finding a blameless victim to condemn, he set to work, thanking God for grace bestowed when he felt empty of inspiration.

There had been two years when he was entirely empty of insights. During his mission in Ethiopia, he had fought just to stay alive. Opposition from the Ethiopian Orthodox community was unexpected, but the violence toward, and kidnapping of, his female parishioners by Al Shabaab was all too expected and abundantly delivered.

His mind drifted away from his sermon draft, and back to those bad years. He had gone there to do penance, but not to lose his life, which nearly happened more times than he cared to recall. His mission was to help build a school for girls, one that would teach education basics but also the words of Christ. And the girls had learned, and thrived, until one by one they had started disappearing. They were not at home sick, watching TV, or texting like their American counterparts. They had been kidnapped to make 'brides,' the euphemism for sex slaves for local Muslim militants. Two years of digging latrines and wells, building walls for classrooms, teaching English, reading skills, and the Bible, had resulted in only two of his students surviving. Two, out of 26. Those two had run from their tiny village to a larger one with a thriving NGO-managed agricultural co-op. Because of their education, they had immediately found paying jobs. The others disappeared without a trace. Their terrified parents knew better than to look for them and also knew they would never see them again. God forgive him, but it was sure good to be back in the US.

As sermon writing finally commenced, Clive Lancaster found himself feeding homemade cream of mushroom soup to his wife. She loved soups of all kinds, and Clive found that he had become something of a master soup chef in response. This one, which included heavy cream and a generous dose of Moscato, was her favorite. After losing her sight, she had slowly lost control of her limbs, and feeding herself was no longer on the menu. This form of MS, Primary Progressive, had attacked the love of his life, his high school sweetheart, the day after she turned 40. A strange weakness seized her for a short time, and she found herself unable to walk. Then she recovered. Believing it to be nothing more than overindulgence in birthday bubbly, she and Clive both dismissed it. Then it happened again the following week and was much worse.

A month later, she was diagnosed. The prognosis was poor. Clive immediately took her on a month-long trip to Costa Rica that they had been planning for the following year. And good thing too. By the time the original trip date would have rolled around, Sharon would have needed to be rolled onto the plane in a wheelchair. Two years after that, she had trouble swallowing anything except soup and could only sit up for a while. Clive didn't mind the caregiving, and they had three other people in the community coming in to help on a regular basis. But saying he didn't mind caregiving was not the same thing as not minding her condition.

Once, she had been the life of the party, and always in a good way, making people laugh with her kind yet pithy wit. She was the one who always convinced her husband that they could put just a little more of their money toward this or that charitable cause. Ironically, one of those causes had been hospice. And to think, it was at one of those charitable events, when Sharon had become too weak for any physical affection, that he had met Tammy.

Randy Miller was just finishing up in court. Things had gone his way, or rather, his client's way. He was feeling once again like a judicial contender. In the courthouse hallway, his cellphone rang.

"Miller."

"Hey buddy, it's Jonesy. Say, have you heard that the body of that bit on the side you had a few years ago has turned up? Hwuuuuwee, did Margo ever find out about her? Cause if she did and now she's dead, well you know what her daddy is going to make of all of that don't…"

Randy disconnected. He no longer felt on top of the world.

Alex O'Connor released a chokehold on a suspect as the man started to turn blue. Ever since the body of that bitch had been found, he'd been in a vicious mood. It was affecting his performance today. His partner glared at him and uttered, "What the hell man, you lookin' for another date with internal affairs?" While the B & E suspect's back was turned, he gave his partner the finger, just out of sight of his body cam. His ex-wife was suing for overdue alimony, he had found out this morning. Attachment of his wages was floating through cyber chat. Alex was now sure that the collective witches of this world were what was wrong with it. That and liberals in general. He cuffed the suspect, a man with a long rap sheet, and none too gently guided the top of his head into the squad car. He hated it, but his partner was right. His life was a steaming pile of dog crap right now and he had to tread carefully. At least he could take comfort in the knowledge that Tammy Splann was well and truly dead and gone, no matter what Duckworth had to say about it.

Back at Bad Aspen, Mike Horvath had arrived in a cloud of dust, pissed as hell. He raced to the pit, seized his wife, looked her over, kissed her hard, and then shook her. His following words were not meant for the ears of folks like Tiffany, so it was a good thing that the Riveras had collected her and were on their way home.

Once the stream of expletives was done about people not keeping their %*$#! noses out of official police business, he turned on Laurelyn with a growled, "You OK?"

"Fine, real, really," she stuttered, never having seen this side of Mike before.

"Good, now get the hell away from this pit. A forensics team is on its way."

"A forensics team? Is that really necessary?"

"Yes, Lor, I think it is." Sandy held up her hand to Mike to forestall an imminent explosion. "See here, the bank has been undercut. That wasn't like that yesterday. Someone not only planted a warning note on a sharpened stake, but they also made sure that when you stood on the edge to retrieve it, the bank would cave in."

Laurelyn paled again. When she examined what was left of the original side of the pit, she could now clearly see shovel marks in the bank.

“Note,” Mike demanded, thrusting out his hand after pulling on gloves.

“Yes dear, but if you ever shake me like that again, you might find a healthy portion of Ex-Lax hidden in your morning oatmeal. Just sayin’.” Her glare gave weight to her words.

Mike decompressed with a groan. His anger had been based in fear, and now seeing that everyone was miraculously unharmed, he apologized. Nancy, taking that as her cue to exit stage left, said, “I think I just remembered a last-minute appointment in Delta.”

“Oh no you don’t, Ms. Clifton. You go up to the house and stay there until I have your statement. Cancel all plans for the rest of the afternoon.”

“Don’t you ‘Ms. Clifton’ me, Officer Horvath, you’re the one that called me here, remember?”

"That call was not to ask you to interfere with a crime scene..."

"Let's stop this, all of us. Mike, honey, we were wrong to come out here again and again. But we have valuable information that you should know, that the authorities didn't have the time to collect. Let's all go to the house and let forensics do their thing." And with that pronouncement Sandy opened her arms wide, fingers wiggling as if herding a flock of unruly geese. Laurelyn, covered in filth and chemicals, was more than happy to be herded, Nancy less so. Mike stood his ground, arms crossed. Nancy barked acidly, "Coming to take my statement or not, OFFICER Horvath?" Three pairs of female eyes bored into him and he gave ground before superior numbers.

"Fine", he growled.

Ten minutes later, as George and William sat in their truck, forensics pulled in. Forensics in the form of one man who happened to be in Hotkiss having lunch with a friend at the Fork it Over. He never left home without his kit and camera, and today was no exception. He'd grabbed the rest of his French Dip sandwich, and headed to his vehicle, leaving his companion to take care of the check. In the parking lot of Bad Aspen, he yanked on his crime scene suit.

George had been sitting in his truck, counseling the distraught William on better ways to advance his religious cause, like being a living testament to the power of Christian principles. William had heard little of this rhetoric in the clink since mostly the teachings had to do with avoiding violating any and all of the Ten Commandments, as well as eschewing the seven deadly sins. The basics were about as much as most cons could digest and retain. The idea of righteous living being its own conversion tactic had never occurred to William, and he was giving it serious consideration. Right up to the time the forensic guy showed up and tromped past their truck in a moon suit. William and George stared at the scene, then at each other, then back at the technician as he made his way down the path past the house and out into the pasture. They had seen Mike's heated arrival, and just thought it had something to do with violating the crime scene, something that George disapproved of but was helpless to prevent.

"Well, OK son, it appears that you may have your first opportunity for providing support and compassion without uttering a word about Jesus. You up for this?" William gulped and nodded, then climbed out of the truck. The two guys could see Mike and the women heading toward the house, and Saffron heading toward them from the chicken house. As he approached the erstwhile sleuths, he saw Laurelyn covered in filth from boots to waist. He began to run. William trailed after.

George enveloped her in his arms. Only then did she begin to cry. Mike stated, "Hey George, excellent timing. There's been a threat. Let's all go up to the house."

Beth, having in earbuds and contentedly listening to music while planting color baskets, was oblivious to all of this. Sandy peeled off from the group and went in to fetch her, reasoning that a threat against one was a threat to all.

Saffron, hearing the word threat, began to lose it, her lower lip trembling. William went to her and put his arm around her shoulders. In a surprising response, certainly to those who knew her well, she relaxed into the embrace of the person she had so recently assaulted. The events of the last two days were taking a toll.

The entire team, including Beth, sat at the kitchen table. William and Saffron had no part to play in the debriefing, so George handed them Saffron's coat and asked her to show William the gazebo and get his opinion on where the trim should be installed. Saffron, knowing herself both dismissed and protected simultaneously, was happy to comply. In companionable silence, the two headed back down the path and out to Saffron's Garden. Although Weston's curiosity was killing him, he stayed in the barn with the newest mom and kids.

Once the older folk were alone, Mike unfolded the piece of paper. By the edges, he held it up for the group to see.

'*Back off or DIE.*' The blocky, crude handwriting, ugly in its execution, was almost as bad as the meaning itself.

"OK Nancy, you first, since you had the least amount of exposure to the scene. Did you notice anything unusual in and around the pit? In addition to the note and the stake?" Three more standardized questions and Mike dismissed her. "Good luck," she mouthed at Laurelyn and Sandy.

"Beth, you next, since you have had the most common sense of all of this lot and mostly stayed away from the area. I know that you have been poking around a bit, though. Tell me everything you've observed, and everyone, absolutely everyone, you have spoken with about this." She admitted to texting George initially, and then the Riveras later. No, she had not seen any suspicious activity or persons hanging around, except perhaps that Deborah person, if she counted as suspicious. That information recorded, Mike cautioned her.

"I think it may be prudent for you to head home for the day. If you are like all of us, you seldom lock your doors or windows, but the next few days at least, I am asking you to do that, out of an abundance of caution." She promised to do so, and knowing Beth's personality and recent experiences with death, he had confidence she would.

"If anything seems off, or anyone seems to be hanging around, don't text me, call me immediately. Head on out now, Sandy can finish up for you, I'm sure."

“Yes, sir, officer, you got it.” With that, she nearly sprinted to her car.

“George, what have you got to say for yourself or these snoops? Any thoughts on this turn of events, lawyerly or otherwise? See anything I should be aware of?”

George was silent for a minute. He knew Mike would wait for a considered response and not pressure him. He finally concluded he could add nothing of value, except, “Not that I can think of, but you might ask Weston, he has been around that area of the pasture more than anyone else.”

“Where is Weston?”

“He just left the goat barn to head up to my place to help Carlos with Shamrock. She seems to be going into labor.”

“Shamrock?”

“My Trakehner mare.”

“Oh.” Mike had no idea what a Trakehner was, and didn’t care to, not being a horseman. “OK, can you ask him to call me when he gets his first chance? George took out his phone and texted Weston with the request. “He probably won’t have a signal, so it could be a little while.”

“Thanks, now if you don’t mind, I’d like to have a word with these two alone.”

George glanced at Laurelyn, then said to Mike, “Sure but could you start with your wife first and give this one here a chance to change her clothes. She is, after all, the intended victim in all this.”

“So we presume, but point taken. Ok, spouse of mine, let’s chat.” Sandy held up her finger in a wait-a-second gesture and retrieved her wine glass from earlier. She filled it to the brim, and only then sat back down at the dining table.

"OK, hombre, bring it on."

She regaled him with stories of sleuthing, with tips and tales from the most unlikely of sources, with photos of receipts, and lastly, with personal conclusions.

"I never thought these words would leave my lips, but I think Deborah is onto something. I think any one of those dudes in her 'murder club,' as she labeled them, are prime suspects. But would a banker or preacher climb into a filthy pit and put a stake in the bottom? A lawyer, maybe, because we all know how they are. A cop? Most certainly. I suggest we look at that note again." Growling, he laid the note out on the table and she took a picture of it.

“This information is not to be shared with anyone else, and that photo not to be shared until it has been logged into evidence. Got it?”

“Ya vol, mein kommandant!”

“Smartass.”

“You know it.”

About then, Laurelyn emerged from her bathroom. She had scrubbed and changed but needed her own large glass of Pinot Grigio. Having secured that refreshment, she plopped down at her table, preparing to be as well-grilled as the sea bass at the Fork It Over.

Chapter 10

"My lovely wife has filled me in on an entirely new world of local crime intelligence. I presume you know everything she does? OK, now, I want to know how someone could access that pit. Is there any way to get there other than from your farm?"

"In a vehicle, no. But someone who wanted to park on the side of the highway and hike down in the dark with a shovel could just jump the fence. Since I haven't installed goat-proof fencing yet, the barbed wire fence along that side is in poor repair, so any able-bodied person should be able to manage it."

"Do you mind if we go up there now? I want to see if there are tracks, fibers, tire prints, and so forth before the forensics guy leaves."

"Yeah, sure. Let's go now."

"Not you Sandy," Mike warned. "You have a greenhouse to close up. And a dinner to fix."

"Fine," she barked, and with wine glass in hand, headed out the patio door.

Physically and emotionally exhausted, Laurelyn was treated to an endearing sight when she least expected it. As she and Mike headed down to the pasture, she saw George standing slightly apart from William and Saffron, who were unexpectedly standing shoulder to shoulder. Each person in turn, gestured or pointed to this or that portion of the gazebo, plainly discussing grand plans for fancification. It was charming in the fading sunlight and caused her to take a deep, much-needed breath. With renewed vigor, she and Mike approached the forensics guy.

“Greetings, Luke. Got any good words for us on who did this?

“Not really Officer Horvath…Mike. Someone was here wearing a man’s size 10 wellies, the kind sold in every hardware store in Colorado. I can confirm that the stake was sharpened quite recently. The soil disturbance evidence would seem to indicate that the excavating was done last night between the hours of 9:00 PM and 11:00 PM.”

Mike stared in amazement at his colleague. “Holy Christmas, Luke, how did you figure that bit out?”

“Elementary, my dear Watson. See this pile of slag over here? This dirt pile is new. So, it came from the pit sometime in the last 24 hours. On the north side of the pile, in these tiny cracks, here and here, there is evidence of frost. That means that it was deposited after the dew point at 12:30 AM this morning. The frost is nearly gone now and certainly will be in another two hours on a day when the climatic conditions are about the same as yesterday. Tomorrow after it snows I would not have been able to tell you this. So anyway, that makes the time of deposition anywhere between sundown to 1:00 AM. But...because the temperature beneath this pile on the original surface is still two degrees warmer than its sides, I can tell that the pile was deposited here while the surface soil still had some warmth from the day's insolation. Sunshine. So, it's reasonable for me to conclude that the pile was deposited no later than 10:00 PM, insulating the surface and standing stationary when frost was deposited two to three hours later. My educated guess, and it's a very reliable one, of course, is that the digger was here well after dark, yet before 11:00 PM last night."

Mike was truly impressed. He turned to Laurelyn. "Did you guys come down here at any point yesterday?"

"Well um, actually, we did."

"Was that pile here then?"

"Not sure, but I really don't think so."

"So if you remember this correctly and the pile wasn't there, then that's additional confirmation that Luke's right and someone was indeed here last night between 9:00 and 11:00. Now think, did you hear anything unusual at all around that time, like a neighbor's dog barking, or...?"

“Or tires squealing away from the scene of the crime. No, sorry, last night was a late birthing, and even though George and I were in the goat barn until very late, I can assure you we were preoccupied and didn’t hear a thing.”

“So you were in the goat barn with the lights all on into the wee hours?”

“Well, yeah, I guess we were. So whoever did this took a risk of being spotted if they entered from the farm side of things. Is that what you’re thinking?”

“Yeah, it is. So now it’s definitely time to walk to the highway fence line and see if there are any signs of a vehicle pulling over there. Luke, could you accompany us in case we need a cast of a tire print?”

“You betcha, wouldn’t miss it.”

The three of them made their way to the highway fence line. There were prints of all kinds, as well as a bit of trash. Despite not being a formal pullout, this portion of the road offered a spectacular view of the West Elks. Shutterbugs had pulled over here, quite illegally, for four decades. But today, there were tire tracks that seemed quite fresh. Mike and Luke conferred. Luke got out a sizable bag of plaster of Paris from his backpack. Taking his water bottle, he formed a paste and applied it to an area with a clear imprint about 14 inches long by nine inches wide.

"OK guys, I got this. I will let you know what the tech boys have to say about this particular make and model of tire. Now shoo, this lady has had enough for one day, and she is wilting on the vine. I know, believe me, because my wife looks just that way when she's had enough excitement for one day."

Mike escorted Laurelyn back to the house. She retrieved her glass of wine and sat down at the table with George. William and Saffron joined them. She poured the rest of the bottle into three glasses. Mike, saluting them and thanking them for their cooperation, gave Laurelyn the same caution as Beth, and departed for his home.

Against Mike's wishes, Sandy had texted Laurelyn a copy of the death threat note. Her phone beeped and she retrieved the image.

"OK, let's see it," George requested. Laurelyn passed her phone over to him. He examined the image and then passed it to William, who shared it with Saffron.

"Um, I think, well, I'm not sure, but I think that handwriting is familiar."

Choruses of, "What?" "No Way!" "Who?" "How?" ensued. The words of the speakers tumbled over one another, quickly overwhelming Saffron. Laurelyn held up her hand to the fellas, hushing them.

"OK, let's go slow here. Please tell us what you meant."

Saffron asked Laurelyn for her phone. She scrolled through the most recent images. She came to the receipts that Nancy had taken a photo of and then forwarded to Laurelyn. She pointed to a signature on one of the credit card slips.

They all examined it in turn. Sure, there was a similarity, but how could anyone be sure? When George asked her, she responded, "Calligraphy."

“Of course!” Laurelyn yelped. “And while you were studying calligraphy to do your wreath labels, you started studying handwriting analysis. By golly guys, there’s a strong possibility she’s right!”

Connecting the signature and the name with a particular person he knew, “I sincerely hope not,” was George’s first response. “If so, I think you better lock all doors and windows for sure, every night. That guy is no one to mess with.”

George’s phone beeped. It was Weston. Shamrock had given birth to a bouncing baby boy, but the new first-time mom seemed confused and distressed. Weston said he would call Mike shortly, but that George and William needed to come home. George stood, embracing Laurelyn and giving her a deep kiss, both because he wanted to, and to piss off his son. He then gave Saffron a light hug and off to the ranch the two men headed.

Laurelyn, checking with Saffron that everything was closed for the night, thanked her for her keen eye and told her she could retire for the night. Feeling anxious about farm security, but also anxious about going outside in the gloom, she dithered to herself. Anxiety about her goats won out, and she grabbed her coat and headlamp. She made a side trip to the chicken and turkey pens and padlocked them, something she seldom did. Only then did she head to the goat barn.

It was such a peaceful scene in the evening illumination that she immediately took a deep breath and began to decompress. There was nothing like baby goats to set one's soul at ease. Belle's triplets, now 48 hours old, bounced around their mother like tiny windup toys on springs, and with about as much purpose and dexterity. Laurelyn knew that in just a week, those same toddlers would be launching themselves onto the large utility cable spools she had set up in the small paddock. The rest of the babies were happily asleep, tummies bulging. One more doe to go, she thought. And then maybe her farm could get on with the business of growing green things. Threats and murders aside, of course. But for now, she soaked in the sight of happy baby goats and their mothers. It gave the phrase 'manger scene' a whole new meaning. She turned the sound on the monitor up a notch. The normal barn sounds of animals getting up and down in their bedding, talking to their young, nursing and even cud-chewing would be audible. It might make for a very poor night's sleep for her, but it was worth it for ultimate peace of mind. She padlocked the door, as well as those of both greenhouses. Only then did she feel some measure of security.

Her concern was warranted. A vehicle parked up on the roadside with its lights off made a hide for an observer. That observer had no interest in the animals, but a great deal of interest in Laurelyn, her actions, and those of her friends. Knowing his message had been received, because of the fresh tracks and new crime scene tape around the pit, he headed up the valley. He hoped the parties warned would take it to heart. The local police were too incompetent to be a threat, but with help from nosy locals...well you just never knew. Especially if one of those locals was an attorney, and one the wife of a local cop.

The vehicle entered the gravel driveway of the Trojan Ranch and turned its lights off. It crept forward because there was a lot of activity up around the small ranch house and its adjacent barns and stables. The watcher didn't know what was going on exactly, but the activity appeared to have everything to do with typical ranch work, and nothing to do with investigating Tammy Splann's demise. After 10 minutes, the watcher slowly backed down in the dark, reassured.

The vehicle then climbed the narrow road to Mesa Splendor. It paused on one side of the vineyard where the house could be observed. The watcher had to exit his hide and get out binoculars. He could see the silhouettes of the couple who resided there. Arm waving and gesturing were apparent, as was turning around in agitation, advancing and retreating. This couple was having a proper row. Could be typical domestic quarreling, or it could have to do with Tammy. He was willing to bet it was the latter. How did one get a cop and his nosy wife to back off? That was indeed going to be sticky, if even possible. He had miscalculated the depth of involvement that Laurelyn chick had with her buddies in this. The number of people on the trail of the killer appeared to be anywhere from four to eight in number, depending on who knew what at this point. That wasn't even counting the cops, and Tammy's ex-husband with his vulturine lawyer. Then there was the public defender, the ME, and district attorney, and God knew who else. No, intimidation of the various parties was no longer an option. Another plan, perhaps subterfuge or a red herring, would be needed if this thing weren't to blow sky-high.

He flipped on the lights and continued on the mesa loop that would eventually lead back down to the valley floor. Angry and anxious, various schemes were examined and discarded. Then suddenly, one stuck. Yes, that might actually work, or at the very least buy time. And right now time was of the essence. The cerebral wheels turned and churned, and at last, a scheme was hatched.

The scene at Mesa Splendor was indeed tense and the topic was that which the watcher'd assumed. It took place after dinner. Sandy had whipped together her unique specialty, Hungry Cop Chops with garlic mashed potatoes on the side. Mike's favorite, extra thick pork chops were stuffed with apple and bacon bits braised together, then blended with Bad Aspen Farm goat cheese and shallots. This was liberally crammed into the centers of those custom meat jackets and the chops seared on both sides with herbs. She topped the whole shebang with red-eye gravy. Not heart-healthy, to be sure. But she had been as bad for Mike's heart today as this dish was sure to be, so no sense sweating the small stuff.

She had plied him with a glass of their own award-winning Tokaj and convinced him to stuff all his words away for a while if he wanted his stuffed chops served promptly. Reluctantly, growling cop-like imprecations, he collapsed in his Lazy Boy chair and let the honeyed juices of this unique wine deflate his angst. All had turned out well, after all, he mused. It was how close someone he considered to be under his particular protection had come to harm. As he sipped, he realized he was feeling guilty. And for what?! He had done nothing wrong. There was no way to know that someone was that interested in a three-year-old crime. Should he have seen it coming? Perhaps. Should Sandy have kept her nose out of things? Certainly. But all of those souls under his wing were actually safe, and he was off duty now. So off-duty be it. He sighed. But he was still going to give his wife a piece of his law enforcement mind. Seeing his color lessen, Sandy topped his glass off again. An hour later, with an excellent meal and beverage under his belt, Mike let her have it, but not with much force.

"You do realize that you girls have come to the attention of some no-good-niks."

Sandy burst out laughing so hard, she wet her pants a little. "No-good-niks is something my grandmother used to say. Oh my God, I cannot believe those words just left your lips." More guffaws.

Mike smiled slightly, then added, "Seriously, Laurelyn could have been maimed or killed if she had landed differently. I blame myself for not seeing this coming. I should have put the pit and the farm itself under surveillance. A fence and a tent over the scene would have been recommended procedure for any city crime, but I let my guard down here and someone I care about nearly paid a price. And it was just a matter of chance that it wasn't you that went down that hole to be impaled. I have been unprofessional, and it will stop this night. I am ordering that field to be closed to all and a fence placed around the pit until the murderer is found."

"You can't do that! Laurelyn needs access to that field to get it ready for this year's hay crop." She did not know that her emphatic gestures about her husband's over-reaction were providing a visual feast for a lurker.

Mike bolted dramatically from his chair. "I can and I will! Get your priorities straight, Sandy. Which is more important, her safety or her crop!"

"You gotta lotta nerve even thinking of asking me that, bucko. Sure, fence the pit, good idea, but excluding the whole field? Imagine if someone said to us we couldn't enter our own vineyards for some unknown duration. It could mean the end of this enterprise and you know it. Besides, she and Weston would probably ignore you anyways."

Mike plopped back into his chair, crestfallen. "Fair point. OK, fence it is, but if you ever tell anyone I yielded to your demands, I'll spank you till you can't sit down."

"Promise?" she retorted with a twinkle. All hostilities concluded, he chased her around the living room for a little while, tickling her. Then they headed to the bedroom.

At Bad Aspen, Saffron got out of bed and examined the note and the receipts again. Laurelyn had sent the pictures to her printer and now she had hard copies to view. She went over everything, got out her hand-writing analysis books, and recalled what she had heard her mom had told Nancy, who had told Laurelyn within Saffron's hearing. She wanted to be really sure, and the more she studied, the more sure she became. A cop. The handwriting belonged to a cop.

Chapter 11

George and William were knackered. It had taken them four hours to get Shamrock settled into motherhood. The little colt had finally nursed well and collapsed with contentment in the fresh shavings in the stall. William had too. He was an emotional wreck and physically exhausted. George covered him with a blanket and let him sleep while he kept watch on the very expensive mare and her even more expensive offspring.

George had just experienced quite the row there in the birthing stall with his own offspring. His attempt to mess with his son's attitudes about Laurelyn, on top of everything else William had endured that day, had proven to be too much for the lad. While scarfing down jerky, granola bars, and strong coffee in Shamrock's stall, William's fatigue got the better of his mouth.

"Dad, I gotta say it. What do you see in Laurelyn? I mean, come on, you've got to be the most eligible bachelor in the entire valley. Trouble seems to follow her wherever she goes and maybe there's a reason for that. I can't believe there aren't some dandy Christian women in this town that would jump at the chance to become a wife and would be much better for you than her."

The silence was deafening. George's color slowly rose, but his lips remained clamped. He knew he would have to say something soon because if he didn't, he would sock his own son right in the eye. Poor parenting, for sure. He took several deep breaths, which William found more alarming than the silence. Only then did visions of being shipped back east for his impertinence occur to him. He started to speak again, to walk back some of his language, but George raised his pointer finger to the sky. Time out. His mouth contorted itself for at least another minute. Finally, he was able to form a civil response.

"Son, I love you, but if you ever, ever, speak words like that to me again, I guarantee you, you will be leaving town on the next Greyhound out of Delta. Do I make my meaning clear enough? Never…ever…again."

"Oh God, Dad, I'm such a screwup, I promise, never again, ever, it's just that…" George held the finger aloft again, and William desisted.

After another minute, George continued, "I know the isolation here must be a drag for a young man. I will compromise with you. Lay off Laurelyn, no, better than that, become downright nice, and once a month, maybe twice, until I get you a vehicle, I will drive you to church on Sunday. I won't go in, at least not for a while, but who knows, maybe one day. Unless you have a different preference, I have heard good things about one pastor and his church in Hotkiss. Seems a lot of young people are going there. Reverend Duckwith, or something like that. Does that seem like a fair trade?" William gulped his gratitude out, nearly strangling himself with relief, and lunged over to give his father a bear hug. His father hugged him back and then held him away, slightly embarrassed.

"Now, how about some more of the coffee, son. It's going to be a long night."

George had watched his son sleep after he dozed off, coffee notwithstanding. Shepherding this troubled young adult was stretching George's reserves to the max and he hoped it would be worth it. He did remember the initial joy of watching his infant son sleep when he had first been born before his mother whisked him away to be with her family back east. He tried to recapture that, and to his surprise, found that a little of that sentiment did glimmer through the haze of past disappointments. He smiled and took a little cat nap himself. The cold woke him up half an hour later. Seeing both dam and son reclining in the bedding at last, he rousted his son.

"Come on kiddo, it's past midnight and we have to be up at 6:00 AM to get ready for those horses coming at 8:00." After reaching their bedrooms, both men fell asleep on top of their beds, clothes, and boots still on. Carlos had stoked the fires before he left, so the house was warm enough for the pipes not to freeze, even if a chill was noticeable, or would be to folks less weary. They slept the sleep of ranchers in spring, which is to say it was deep and all too short. Calving had begun a week ago and would not conclude for the next several weeks. It was a hard life, but it had its compensations. George did not begrudge it.

In Grand Junction, there was an individual drinking alone in a dingy bar who did begrudge his life, very much so. The big red-headed deputy sherriff nursed a boilermaker and brooded. It seemed like every person on the planet had a better deal than he did. After that morning's bust, his outlook had brightened. But a warning from the sherriff about police brutality had tanked his mood, one of incipient and constant anger, to the next level down, anger and self-pity. But it was OK, he actually preferred drinking alone than with those pompous knobs at the Fork It Over. The crap he'd had to endure, especially from that preacher! But he had a plan, a plan to even the score, level the playing field, serve a little backhanded justice. His thoughts caused him a bit of glee, and he smiled a crooked, nicotine-stained smile. He called for another bourbon, and mused, *yeah, you think you all know each other's secrets, but ya don't, not by a long shot.*

"Sorry Alex, closing time. Even if it wasn't, I'm cuttin' you off. Call a cab, get some sleep. See you next time."

He thought about lunging for the barkeep, despite the fact that the big biker outweighed him by 40 pounds. But Spike could and definitely would call the cops on a sloppy, hostile drunk, even if the drunk was one of their own. No favors granted, so no suspended liquor license, or worse. And Spike knew from worse, being an ex-con and all himself. So grudgingly, Alex paid his considerable tab and staggered out the door. Spike dialed a cabby who he knew was waiting just around the corner to come get his last fare of the night.

Dawn broke snowy, as predicted. Weston cursed the events that had allowed the beautiful weather to depart without him being able to finish plowing. Despite the persistent drought, he hoped this snowfall would be light and dry, so he could get back at it tomorrow. He worked exclusively for George and Laurelyn these days. He had sold all of his cattle to George so he didn't have to rush home to feed and water them daily. Besides, things were going just swell with Sally, and he wanted to be able to have a social life now and then. So, although Beth and Saffron were a tremendous help to Laurelyn, he felt himself to be the true farm manager and was very loyal to this misperception.

Laurelyn was, of course, the owner/manager in fact, and not just in theory. She knew Weston felt this way, but he was so skilled and reliable, not to mention affordable, she humored his over-investment in her operation. Better too much stakeholding, than too little, she reasoned.

He loved working for George as well, although William's presence made the situation a bit murky, and stole some of Weston's thunder. But he figured he just had to outwait him, since the kid did not seem like a long-term rancher, at least in his humble opinion. He admired George, though, for doing the right thing by his son. Weston's own folks would never be nominated for parents of the year, but they had certainly been better than what George had endured.

He definitely looked to George as a role model. But when it came to expressing manly virtues of integrity and leadership, he preferred his role at Bad Aspen. He reasonably concluded that it looked like it would be a barn cleaning day for him, since greenhouses were the domain of the fairer sex. He would take a snow day into town as well for more shavings and stop in to see Sally. But first, he took it upon himself to stop in to clean the barn and make sure all baby goats and moms were well.

It was a good thing he did, because one of the newborn bucklings had managed to get his big head stuck in the stall fencing. He must not have been that way long, because he wasn't crying and wasn't cold, yet. He would have been soon though, as his anxious mother had nearly licked him bald in her worry, and he was damp and would have chilled soon if not for Weston, who wondered why Laurelyn hadn't seen this on the monitor up at the house.

Laurelyn and Saffron were hunched over printouts. Saffron had said nothing until this morning, knowing how much Laurelyn needed her rest. But while Laurelyn was hoisting her second mug of expresso, Saffron came from her room early with tidings of her discovery.

"So before I share this with Sandy, are you absolutely sure? Show me again exactly the evidence you have that leads you to believe this is the same guy. Looks different to me." Saffron pointed out similarities in word slant, the fact that in both cases the words trended downward on the paper, similarities between the 'o' in the signature and the 'o' in 'off' from the note.

"See how the words slant to the right in both cases and go downward as well. The signature on the credit card slip indicates a sad and angry person who is relaxed. The writing on the note is from the same writer but in an angry and agitated mood. Here, this book has some examples of how writing changes with mood, but still retains certain features." Retains certain features. This was delivered in a professional manner that would never have been possible only six months ago. Laurelyn had an anxious moment when she realized that one day Saffron might fly the coop because she didn't need mentoring and parenting any longer.

It was at this time that Weston discovered the little boy's predicament and freed him. He spruced up the barn bedding and decided he couldn't complete the task without those fresh shavings. On his way up to the house, he noticed the female huddle at the table. Curiosity, his middle name, began to poke its ungainly head up. He knocked briefly and then entered.

"Hey there gals, what's up? Crime-solving? Don't give me those looks. Anyway, I was just in the barn and a baby goat was stuck, but all good now. Since it's snowing, I'm gonna head into town for shavings. Saffron, Sally wants $10.00 worth of greens this week if you can manage it. Can you bag those for me before I head out?"

Both women bristled at his announcement, not request, but knowing his heart was in the right place, they simmered down.

"Just doing a little hand-writing analysis for fun, nothing much really," Laurelyn lied. "Thanks for checking the barn. Shavings are a good idea, I bet we're low. I was going to get them tomorrow but today is even better. Saffron, could you get those greens for him?" Once she regained control of her temper, Saffron was glad to comply, because that $10 went right into her personal fund.

Laurelyn desisted in her handwriting examination, feeling convinced, pretty much, by Saffron's detailed explanation and examples from her books. Curiouser and curiouser. Why would an officer of the law want to harm any of the farm help? He should welcome a chance for inter-county and inter-agency cooperation. But as part of the 'murder club', he plainly had something to hide. Like the fact that he was the killer, maybe? Sandy needed to know about this immediately. But first, goats.

Nine baby goats were bouncing around the mams. A more joyful sight Laurelyn could not have wished for. She urged Belle up onto the milking stand. There would be a point just before weaning when she would have no milk to spare. Her triplets would make sure of that. But right now, her udders needed some relief, and the farm needed milk. Belle enjoyed the relief and massage, and the dairy goat ration that went along with this procedure. After she filtered the milk and set it to chill, Laurelyn texted Sandy with Saffron's analysis.

"Chris Chringle on a cracker!... to quote Sally. You think he could be the killer? The cop in the murder club?

Yeah, George's right, if it is this dude, a crooked deputy is no joke. I'm calling Mike right now to let him know, cause this is huge, and scary. I want to make sure that not only is the fence is put in place today, but that there's extra police presence installed as well. Of the good kind." For once, Laurelyn did not argue with that plan. Especially if Weston wasn't going to be around a lot today, and George was occupied with new horses. Which he was.

She hung up. She called George, but it went to voicemail. She left a message with an update, hoping he would get it soon. Then she called Beth. She could tell the news frightened her, but Beth insisted on coming down to the farm anyway, to continue her color baskets and plant some more hothouse herbs like thyme and rosemary. "OK, more people are probably safer, but lock yourself in while you're in there. We don't want you pushing up daisies, just planting them."

Beth was too nervous to appreciate the humor but did appreciate the suggestion.

Sandy's phone text buzzed. Assuming it was George she glanced at the phone but saw an unfamiliar number. A B&B booking inquiry perhaps? She retrieved the text, and was dumbfounded. Someone had sent a newspaper clipping, an old one. It seemed to have to do with Lawrence Duckworth. The subject line of the text was blank, but hand printed in the margins of the article were the words, '*Plenty to hide, and plenty of experience with guerillas in Africa. Knows deadfall pits.*'

Chapter 12

Sandy wondered for a minute if the sender meant gorillas, but as the implications sunk in, she realized he or she had written just exactly what they meant. Someone besides Mike, the forensics guy, the ME, and…she realized that the list of people who knew about the sharpened stake must surely be numbered almost to 15 by now. More, if anyone had told Margaret Stanley. No point in 'process of elimination' thinking, none at all. She enlarged the image of the news clipping and read the content. A religious sex scandal! My oh my, reverend. And now an enemy wishes to implicate you? The deputy? Probably. But why this reverend guy? Were they enemies? Seemed unlikely if Deborah had seen them all dining together in relative harmony. Did he truly have something to hide where Tammy was concerned and one of the other 'murder club' members knew it? Someone besides the deputy? Or simply a concerned citizen wishing to finger the guilty reverend? Lawyer, banker, Indian chief? Husband, high school math teacher, or totally new unknown subject? Knowing she absolutely had to lay out drip irrigation today, she hustled out to do that. Snow showers mixed with sun would be no hindrance to that task, and the work might help her to sweep out some cobwebs and find insights. She did not pass the text on to anyone just yet, wanting to think about it. She did text Mike to say that there may be an issue with a dirty cop and could extra security for the farm be managed immediately? Please? Then on with the wellies and out to the vines.

An hour later, the lurker was displeased to see Delta county's police presence in evidence at Bad Aspen Farm. Another deputy sheriff's SUV was parked in the driveway. The lurker had to skedaddle back to his home turf but took time to drive by the church that housed the congregation of the Rev. Duckworth. No cops, none. Huh, not good. He concluded that his message may not have been received and passed on to the authorities. He made a vow to check later that evening. Perhaps an additional nudge in the right direction was called for.

It was Simon's day off, and he and Milton sat sipping Ethiopian coffee, one of Milton's housemate's specialties. Dorcas had started importing and roasting beans grown at a plantation close to her home village. The results had been incredible.

"A beautiful day, made even more so by Eddel the Gazelle." Milton quipped at his daughter. The honey-skinned child performed a pirouette that made Simon clap. They were an odd family, but an astonishingly harmonious and functional one. Milton and Simon were a couple, Dorcas, the mother of Milton's child, was single and not-dating, and Eddel was cherished, guided, and nurtured by all three. Dorcas was busy in the kitchen at that moment, so Simon and Milton were free to gossip about the current events. Once Eddel left the enclosed sunroom that Milton and Simon had built, they were able to be frank in their chatter about the murder.

"So what new juicy developments can you titillate us with this morning, my dear?"

"Not too much, really. I think it's common gossip by now that some creep put a stake in the bottom of the burial pit and Laurelyn barely escaped serious injury. For a woman who left several lovers a goodbye note three years ago, she certainly has made her presence known in a more contemporary way.

"What note?" Milton enquired.

"Oh didn't I tell you, the deceased left identical *Dear John* notes to a number of former lovers. McGinty, bloodhound that he is, has been in contact with other upstanding members of the community that were, how shall I say this, enamored of the lady's charms. He canvassed the suspected paramours, those exposed by the husband, to see if they too had received such notes. Turns out, one of them admitted they had, the bank president. He was all too happy to provide it to McGinty to assure him of his ignorance in the matter of her disappearance. I believe that one has now been logged into evidence, but not before McGinty received a photo of it. McGinty opined that to his untrained eye, the signatures seemed to match. All other supposed lovers have made no comment. But the banker says he knows for a fact they all got one."

"Do tell." Milton was silent in thought for several minutes.

"What's up, Buttercup?" Simon finally prodded.

"You don't happen to have a copy of both of those notes, do you?"

"I have the first one from Jack, but not the one from the banker."

"Can you get a copy of that one as well? And is it legal to do so?"

"It's highly irregular, but it isn't illegal. Why, what are you thinking?"

"I'm probably being a Nosy Parker, but why a letter to an affair partner? Who does that? I mean, I find it rather curious. Now people just ghost each other."

"Now that you mention it, that is strange. Not impossible, but a bit odd for sure. But you're McGinty's boy, or visa versa, why don't you ask him for a copy?"

"I believe I will do just that." And pulling out his phone, he did just that. He received a text with an image 10 minutes later with a request for any insights Milton might have. McGinty was just leaving a mimosa laden power brunch in Denver, one brimming with schmoozing and the promise of future lucrative legal work. He was in a bit of a hurry, and so just fired off the text without thinking. He was a little overwhelmed with work for the next several hours, and so hoped nothing, like the arrest of his client, was going to come down the pike. In the meantime, any help was good help.

Simon and Milton held their phones next to each other. Comparing notes, literally, they discovered that the closures, *Yours, Tammy*, were identical. They looked at each other and Milton queried, "Again, I ask you, who does that? I'm all for efficiency in communication, but identical kiss-off signatures? Who says, '*Yours*', to a soon-to-be ex-husband? No, something here is fishy. I know who might be able to help. Let's pop up to the farm today and ask Saffron if these signatures are from the same person. From what I can tell about her hand-writing analysis hobby, she's pretty good, and works for free." They toasted to that pleasant solution, clinking mugs.

"Take Eddel with you, you lazy loafers."

"Yes mistress Dorcas, I live to serve." Hearing this, Eddel squealed with delight and ran off, to change from ballerina to farmhand. Helping Saffron could sometimes be a dirty business.

The snow was departing and the day would warm to a pleasant 40 degrees. The greenhouses were already up to 55 degrees when Laurelyn entered. One last doe to give birth and the season's kidding dramas would be over. As Laurelyn began attending to some chiltipin accounting, organizing orders into first come first served, and getting ready to ship seeds, a couple of workmen were seen tromping down the path from the parking lot to the pasture. She had not seen the law enforcement presence stationed in her driveway and was puzzled. But when she saw that both workmen were carrying three temporary fence sections each, she knew what must surely be up, although Sandy had forgotten to tell her. She called her and received a breathless greeting.

"Sandy Hor...Horvath," pant, "How may I help you?"

"Becoming formal with each other now, Sandy Hor?"

"Oh, I'm out in the vineyard and the sun came out and it's really glary and I couldn't see the number on my screen, and oh shut up you. Anyway, you guys all good down there? Did the cop come by to check on you?"

"Well no one came to the door, but since I can't see the parking lot from here, someone could be parked there. And two guys with fencing just walked by. Know anything about that?"

"Ah, Criminy! I forgot to tell you. I guess because Mike and I sorta got into it last night, I subconsciously blocked it out. Anyway, he wanted to exclude the entire pasture from trespassers, and I convinced him that just fencing the pit to keep all of us looky-loos at bay should be enough. He insisted on the police presence for now, and I definitely concurred. Don't argue, not after what nearly happened."

"Who's arguing? I said before I thought it was a good idea, especially the security presence. I have Saffron and Beth to think about. I was just puzzled for a bit…all cleared up now."

"Well now that's cleared, let me muddy the waters again just for grins. I'm going to text you something I received about three hours ago. See what you think and get back to me."

"Yes mam!"

The text and image received, Laurelyn was freshly alarmed. Perhaps there was no honor among thieves or murderers. The plot thickens. A falling out? A red herring? A genuine tip? Laurelyn's thoughts ran almost parallel to Sandy's. She sure wished she could run all this past Bernie, but alas, still no contact with her. She needed to get her seed orders in order, and so reluctantly put those thoughts aside, just as Sandy had done. She identified five orders that needed to go out today. She decided to run them up to Saffron and have her attend to them. Saffron didn't seem nearly as weirded out by the current events as Beth was, so she thought she would show her the text and get her thoughts. She grabbed enough seeds for Saffron to also fill 10 orders for tomorrow, along with purchaser info.

Saffron was in the craft room making wreath bases out of willow wands. She stopped immediately and took the materials for shipping from Laurelyn.

"Before you get started, check this out."

As she read, her eyes grew wide. She began to smile a little bit, which was the last thing Laurelyn expected. She looked up, and stated plainly, "The note writer."

"Who is, the pastor?"

"No, the sender of this. The sender is the note writer, the one left in the pit." And with that, she took her assigned task into the craft room and started labeling envelopes in her beautiful calligraphy. A tiny hand-written note thanking them for their high-elevation patented chiltipin seed order would be slipped into each packet as well along with a request to leave a review on the farm's website. Watching Saffron set to work,

Laurelyn slid her phone into her overalls pocket and shaking her head, left her young worker to it. No confounding suppositions had clouded Saffron's cognitions. She had plainly examined everything swiftly and dispassionately and come to the most likely conclusion. Laurelyn was beginning to wonder at this point if Saffron might actually be intellectually gifted, and not just artistically inclined. Laurelyn was inclined to trust her insight.

So, the cop had presumably left the pitfall note, and then for some reason, sent this. OK, but did that mean that the deputy was the killer and trying to frame someone else? It seemed like the most parsimonious explanation. Not the only one, but certainly the most likely.

Now Laurelyn really wanted to hear firsthand from Deborah the conversation between members of the murder club. Money had been mentioned, but Nancy said some parts of the conversation had been uttered in lowered voices. Expressions of relief at the victim's death had been loud enough to hear every word, but the bits about money were hushed. Deborah had definitely heard one of them exclaim that '*she was into him for $2,000*,' but she didn't know which of the diners had said it. If Tammy had gotten loans from these men, would they be all that happy to see her go to her heavenly reward? Wouldn't they want their money first? Had they gotten it? But $2,000 lost over a bad loan was surely no big deal. Not worth killing over. But what if it was the opposite, what if they owed her money? She did seem to have some. Did the murder club men have a common gambling problem? Cocaine? Still, the amount of money was peanuts. A puzzlement.

So many thoughts began chasing themselves around in her brain, it started to feel like Lewis Carroll's jolly caucus race. It was right then that Milton drove up. Laurelyn groaned a bit, wondering if she was ever going to be able to get farm work done. No matter, Saffron would think of something to keep Eddel busy. She always did. And she did owe Simon for some light espionage, that was certain. So, time to pay up.

As expected, Saffron took charge of Eddel immediately. It turned out she felt the chicken nest boxes were in dire need of being refreshed, to prepare for the coming egg season. So, they let the chickens out into the yard, a rare treat for those ladies at this time of the year, and gave the nest boxes the same treatment that William had given the Trojan Ranch horse stalls just the day before. George had helped Saffron to find the fun in such chores when done together with a partner, and now she used those skills with Eddel. Silliness, storytelling, and other work-party tools were generously utilized. The result was six spectacularly clean nest boxes and one tired but happy child.

While all of this was taking place, the adults sipped wine on the now sunny patio and admired the sparkling, snowcapped view of the West Elks. Laurelyn showed the men the text from the anonymous tipster. She had assurance that Sandy had finally sent it on to Mike. The cop in the farm parking lot had needed a lunch break, and so, after introducing himself to Laurelyn and crew, had excused himself for an hour. She felt that with the two men, Beth, and Saffron all present, a killer would have to be insane to attempt something in broad daylight. Unless of course, he wasn't sane. But Laurelyn pushed those thoughts away.

The men conferred and discussed and speculated and proposed. They were having way too much fun with this. After ½ hour of this, she decided to give them a job. She took them down to the goat barn to complete the daily cleaning since Weston had returned with the shavings. She then returned to the house to launder the birthing towels in preparation for the last blessed event. Which hopefully would be blessed after all the trauma of these past few days.

As she was putting towels into the dryer, her phone rang with that same San Diego number. Knowing who it was and that now she truly wanted to talk to this person, she answered promptly.

"Hello Deborah, how are you today?"

"Oh, Uh, I'm fine, really fine. How are you?"

"Quite well," the stilted, and socially correct greetings continued.

"Oh that's good, that's good. I'm calling to see if I can make an appointment to get together with Saffron. It doesn't have to be there, but it could be, I mean, if that's OK? Or we could do lunch. Me and her, I mean. You know. Can you ask her for me?"

"I will certainly let her know. But now I have a favor to ask you. I have heard third-hand about the conversation at the Fork It Over. There has been a further incident here, no it's OK, no one has been hurt, it's OK really! Just hear me out. Someone has been sneaking around and I think I really need to hear firsthand what you heard. It could be in everyone's best interest. And also, at some point, the police will come knocking to get a statement from you as well. No need to worry, just routine, but you will probably need to testify in court when the killer is apprehended and tried. No, don't cry. Pull yourself together, there you go. Now take a deep breath, or two, and tell me what you heard."

With many gasps, half sobs, gulps, and stammers, Deborah finally got out the entire tale. Laurelyn was glad she had persevered because there were a couple of important tidbits that Nancy had not relayed. Deborah had no further information concerning money changing hands than before. But she did hear something about notes. All of them had raised their voices for a short while *on the subject of notes*. Nancy had not taken note of the importance of the notes. Deborah was certain that all of them had expressed receiving them, and they all seemed to say the same thing. So anyone denying receiving one when asked by the police… was a liar.

Full of thought, Laurelyn rang off and dutifully went in search of Saffron. Seeing that the nest boxes were cleaned, scraped, bleached, and full of aromatic pine shavings, she felt better about farm chores for the nonce. So, not seeing the girls down in Saffron's garden, she decided to check the goat barn before checking the house. Good thing she did, because the girls had discovered the joys of throwing goat raisins at each other and at the men. Not much work getting done at all, but some much-needed laughter taking place. She smiled quietly and then laughed out loud when four pairs of sheepish eyes spotted her.

"I see what's going on here. And for your sins you four will take all baby goats except these two here, out into the sunlight. Lead their moms out with them, and don't let Titan and Thor out. Saffron, can I see you for just a minute? Hang on guys, she'll be right back."

She relayed the message from the young woman's biological mother. Saffron took it with better grace than previously. Not having been shocked by the woman's unexpected physical presence made all the difference. She looked down and muttered, "Can I get back to her later?"

"Of course you can! All of that is entirely up to you. I am just the messenger."

Saffron, relieved at getting no pressure, scooted off to lead the does out, while the rest of the crew scooped up the baby goats. Seeing Beth coming down the path, she again had a renewed sense of farm progress. She grabbed a half-empty bottle of Mesa Splendor white table wine from the milking room fridge and three dixie cups. She sat on a tree round next to Milton.

The three of them continued their hen party about the investigation. Laurelyn now had something very interesting to share. The goat moms lay down in the peach fuzz grass and let the baby goats jump all over them. Belle, needing a rest, went to lay down on the old straw heaped to one side of the fence for that purpose. Within minutes, she was asleep. She was getting a little old for birthing and might need to be retired soon, Laurelyn realized. But those husbandry considerations would have to wait.

"So what you're saying is that all these guys were sleeping with the victim, and they all got identical letters on the same day telling them she was leaving town. What a courteous blackmailer!" Simon quipped.

"What did you say?" Laurelyn probed. "Blackmailer? You think that was the deal?"

"But of course, mon ami. Nothing else makes any sense. I thought you had already realized that! I happen to know that all of the men at the table are or were married at the time of the affair. Except the pastor. So yeah, affair. Oh come on, Laurelyn, have you been a mushroom all this time? Kept in the dark and covered with manure?"

Laurelyn literally smacked herself on the forehead with her free hand, which caused the wine in the other hand to splash a little. How could she have been so blind? Tammy was blackmailing them on a monthly basis for the amounts that Deborah overheard. And now she knew why the reverend was there. He too was being blackmailed, not for a dalliance with Tammy, perhaps, but because of past indiscretions. Hence the incriminating news clipping. That gave every last one of them a motive. A 'murder club' indeed! Deborah's instincts had been correct.

She turned to Milton. "Is it legal to share our theories with McGinty? This does cast probable cause on several others and reasonable doubt on the husband."

Milton replied, "I don't know if it's legal or not, not being an attorney and all, just a humble entrepreneur." Then both pairs of eyes turned to Simon. The target of their attention rose and seeing that he needed to be out of earshot, with resignation went to supervise the baby goat romp.

"Simon, while you're out there, can you let Thor and Titan out and keep an eye on them. They won't hurt the babies, but their moms may take exception. One never knows in the first couple of days of herd mingling."

Simon blanched a little. He had no experience with animals. Thor and Titan were huge, with sweeping horns. But he decided to man up and do as requested while maintaining plausible deniability.

He let the big fellas out. This was not their first baby goat intro, and they tiptoed forward, keeping an eye out for mad does. The first-time freshener, who'd had such a difficult birth with her twins, immediately charged them. They headed back to their section of the barn like torpedoes, shooting divots of tender sod in their wake.

*Pussi*es, Laurelyn observed. Then for the next few minutes, she and Milton corresponded with McGinty, who was over the moon at the news. Easiest defense he'd ever prepared. And although the killer was still unknown and still at large, well, that wasn't his concern, now was it?

"There, civic duty done. Now, Mike Horvath really does need to know about all of this, really." Laurelyn, hating the tongue-lashing that was undoubtedly awaiting, reluctantly concurred. "But I want you here! Just so there isn't a second body...mine!"

Laurelyn tried getting the big boys out again. She put the nervous doe and her babies away and encouraged the wethers to interact with the others. Success! Ahhh, peace in the valley. The rest of the socialization could wait till tomorrow. In the meantime, most of the herd got to enjoy a sunny afternoon together.

The killer was troubled in his mind. Efforts to muddy the waters, efforts to confuse and intimidate were going exactly nowhere. If only there was just a little more time. If there was any justice in this world at all, there would be time to do what he had to do.

After feeding the barn cat, Myrtle, Saffron took Eddel up to the house. She was starting to run out of ideas for entertaining Eddel when she spied Sally Handy's yearbook. She thought it might be fun to show Eddel a peek inside the minds of high school students. She opened the pages and they exclaimed over funny hairdos and sappy sentiments. "Oh and here's someone I think you know," she said, turning to Sally's senior photo. "This is the lady that owns the Merry Clary. Well, owns it with Beth. This is what she looked like five years ago. Gosh, she hasn't changed much at all."

Then Saffron frowned. She was seeing something that didn't add up. Something that needed research and scrutiny when not babysitting. She took Eddel into the craft room and showed her how to bend the willows into hoops. She knew she would have to redo them, but it kept the little girl busy and entertained. Finally, Laurelyn and the two men entered the dining room. Saffron nodded to Laurelyn and then gestured with the top of her head to the craft room. Eddel could be seen masterfully bending her materials into shape while rubbing her eyes in tiredness.

"Simon says it's time to go home, Gazelle. Simon says...let's go!" It was a private game that had developed over the last couple of months. Simon always kept a

Tootsie roll in a shirt pocket. It was the prize for winning the 'Simon says' game. With a frowzy grin, she slumped off her stool and sprinted the few feet to his side, taking his hand.

"Well done! You win! Why don't you eat that in the car on the way home."

Daddy Milton promised to be home shortly. Laurelyn would take him as soon as was practical.

With the two of them gone, Saffron gathered all the materials she believed she would need. Copies of the notes, printed out, copies of receipts, also printed, and the yearbook were all spread out on the dining table.

"Here is something I don't understand. Look, here is Tammy's best wishes for a happy life for Sally. See how she signs it. And here are some others. She says, "Love and Hugs, Besties Forever, and there's plenty of XOXOXO. Nowhere does she write, 'Yours.'"

Laurelyn, skeptical that this was meaningful in any way, said, "Lots of people change their greetings and closures as they mature. Society and the workplace often demand it."

"Maybe, but there's this. Here are Tammy's yearbook signatures over and over. Compare them with the note to the husband and the note to the banker. The signatures between notes are the same. But look at the note's signatures, and the yearbook signatures."

Laurelyn did so. To her eye, they seemed to match. Nothing seemed all that odd, not really.

With certainty, Saffron added, "They're not the same. Tammy did not write those notes."

Chapter 13

Milton and Laurelyn stared and stared. They plainly weren't getting it. Saffron, in an uncharacteristically theatrical way, sighed. She went and got out her handwriting analysis books, again. She patiently pointed out that the slant was all wrong. Tammy's handwriting went ever so slightly up, indicating an optimistic outlook on life. And in many cases in the yearbook, Tammy had added a flourish to the first letters of her first and last name so exaggerated that the right arm of the 'T' sailed all the way over the rest of the letters and touched the 'B' in Bedecker.

Milton and Laurelyn looked at each other. Now they could see the differences. The notes were very similar to the yearbook, except for one thing. They were too correct. Emotionless. Leaving a husband or lover was an emotional experience, whatever those emotions might be. The note signatures were perfectly level, correct, and robotic.

"Do you have a theory?" Milton asked as he formed his own.

"Well, I think a good forgery would have been less perfect, more human. I think this is computer generated." Laurelyn pronounced.

"My thinking precisely," he concurred.

"How? Why is obvious, to presumably protect the true note writer from discovery. But why not just use one of her own signatures over and over?"

"Maybe there was no clean image." Laurelyn mused. "Plus, as I'm sure Saffron will concur, people's signatures surely vary a little, depending upon their moods or even their health on any given day. And by the way, my dear, you are even more talented at handwriting analysis than you are at babysitting, and that is saying something." Saffron actually blushed.

"But the same unvarying signature copied over and over could arouse suspicion, as these now have," Milton added. Saffron nodded in agreement and added, "Yes, although it's really hard for investigators to determine who the forger is if they attempt to recreate someone else's signature. The forgery may be easily spotted as a fake, but it's really rare for the police to be certain who the forger is. But maybe the author of these notes didn't know that and feared a handwritten fake signature could be traced back to them."

"You know where a rich source of clean signatures might be found? Checks." Laurelyn offered. "Sure, some might have messy backgrounds with puppies and rainbows on them, but these days with so many people depositing checks digitally, someone with access could grab those images and make a perfect amalgamation of them. And who among the murder club has access to checks. I count three. The husband, the lawyer, and the banker. The deputy, unlikely. The pastor, sure if she tithed by check, but she wasn't a member of his congregation, so I think we can rule him out. So one of three of the murder club is off the hook for the notes that Tammy supposedly wrote. But what about the sharpened stake and the note in the pit? What about the actual killing?"

"Could the murder club be just that, a club? With each member having a role to play in her disappearance? And maybe we're letting the husband off too easily. I think you're right Laurelyn, we need to put Jack Splann back up on the murder board," Milton mused.

"Now we know who watches crime shows. Maybe we called McGinty too soon."

At that moment a fist thundered on the patio door. *Here we go*, thought Laurelyn. *Angry cop time*. And she wasn't wrong.

Mike was so tomato red, Laurelyn was sure he was going to have a coronary right then.

"First, is everyone OK here? All present and accounted for?"

Laurelyn assured him that everyone, right down to Silver-laced Polish hens, were all just swell.

“Good, because that same set of tire tracks was found in a different spot on the shoulder of the road. Made last night. You are definitely being watched.” Laurelyn’s gasp mollified him somewhat, but not much.

“Oh my God, Mike, it’s so good that you’re here. And we have so much to tell you. Really. But do sit down, you’re frightening the children.” Both Milton and Saffron had gone a bit pale.

“Fine,” he growled in a tone Laurelyn thought she was hearing all too often these days. He crashed into a chair and said, “So, what is it?”

Laurelyn was going to have Saffron explain it, but seeing that she was about to bolt, throw up, pass out, or all of the above at once, she took the helm.

The next half hour was tense, alright. But despite his righteous outrage, Mike found the evidence they had garnered so fascinating, he almost forgot to be angry. Saffron, who had issues with angry men, was able to decompress, likewise Milton, who had issues with Mike.

"And just when I was thinking of arresting the husband on suspicion. Well, this throws the crap into the blender. Still, I believe I have warrant to get a warrant to seize some digital records. Or rather the detective on this case will have once I give her this. And an interview with that waitress, I mean, Deborah Johnson," he corrected," is certainly in order. I'm forwarding all of this information to the detective. In the meantime, I am going to stay parked in your driveway. The Delta deputy sheriff has been called away to other duties, and now you're stuck with me, so don't try any funny business."

Was that humor? Laurelyn was hopeful it was as she beseeched him, "But Milton has to get home, and I was going to take him."

"Nope, he's staying put unless Dorcas or Simon want to come fetch him."

Mike left to ensconce himself back in his department SUV. From there, he used all of the modern equipment at his disposal to distribute this bit of embarrassingly relevant amateur sleuth information. If only his department was better funded! On the other hand, if they were better funded, they might want to put him on full time, and then he couldn't run a winery and…"

Sandy, having been apprised of the situation, and having already made his favorite dinner the night before, decided on emergency measures. Favorite cop on stakeout Hungarian dessert. Palacsinte. She began to whip up a batch of those stuffed crepes. Without the sugary topping, that if scattered on a cop's shirt might impair his dignity, the fluffy tubular crepes made from his grandmother's recipe and altered just a tad, was the perfect cop finger food. Cottage cheese added to the filling even provided some protein. Local Peony Creek dried cherries gave it added zing. She whipped out her crepe pan and set to it with a will. Half an hour later, she delivered the care package, hoping the sweets would sweeten his mood. They did. That plus a thermos of gourmet blend coffee got him through the next few hours just fine.

An unmarked police cruiser pulled into the parking lot of the church shepherded by the Reverend

Duckworth. Lawrence was in his study, waiting for a parishioner to arrive. A morbidly obese woman was having marital issues due to the effect her weight was having on her health and wished to consult with the pastor as to what could be done, being as she had this terrible sweet tooth and all. Lawrence was spending a moment in prayer, hoping to guide and comfort this woman, without giving either offense or false hope. The knock on his door seemed to presage the woman's arrival. He just hoped she could get through the narrow door with dignity intact. At his invitation, the door opened and to his shock, the visitor was dressed in a suit and tie. She was five foot nine easily, and athletic of build. She displayed her badge and requested a few moments of his time. Lawrence, feeling his past catching up to him, deflated into his ergonomic chair and with a despairing hand, ushered her into a seat in front of his desk. When the expected client attempted to enter the doorway by turning sideways, the detective flashed her badge and announced, "Sorry mam, police business." The woman squished her way back out, and Duckworth squished further into his chair. It seemed that the precinct had received an anonymous tip that Rev. Duckworth may have been involved in the murder of Tammy Splann.

An emergency meeting of the Peony Creek investigative team had been called at the same time that Duckworth was having a tense conversation with a detective. Present were Laurelyn, Sandy, Beth, and George, engaged in funny business. Milton had been picked up by Dorcas and was now home. William and Saffron were ensconced in the goat barn.

At the end of the meeting, Laurelyn had developed a keen urge to open a new account for the farm at the First Bank of Peony Creek. Beth found that she wanted to replace her current attorney with a new one. George found that he needed to vet the church that William would be attending. Sandy was feeling that the current tax management situation for the vineyard had become overwhelming and a professional accountant was needed. All of them would be attending to these tasks first thing in the morning.

Sandy's additional job would be to coordinate all findings and communications, provide central intelligence for all investigative efforts, plus all farm, vineyard, and ranch activities, so no critical needs on the home fronts went unaddressed.

That decided, emails were sent to their prospective targets to arrange meetings. Luck was finally with them because all had their assignments confirmed between 9:00 AM and 10:00 AM the following day.

William and Saffron stood shoulder to shoulder with arms draped over the top railing of the goat paddock. Saffron had led the fractious mother and her babies back out into the pasture to mingle with Thor and Titan. She had swung her head at them, and they had backed off, but just a little. Eventually, one of her own kids bounced up to the big boys and danced all around him. The watchers could almost see the mother heave a resigned sigh and roll her eyes before yielding. The watchers laughed with released tension and genuine humor. Saffron usually laughed with George the most, but in his absence, William would have to do. They began to chat about farm and ranch life, comparing the differences in their tasks and the expectations of their respective bosses. Camaraderie grew. Finally, William could take it no more and spoke his mind.

"I really like you, Saffron. You're talented, smart, and fun. I would like to help you with your garden project whenever my dad will let me off from ranch work. I'll have my own truck in a couple of months anyway, so I won't have to depend on him or Weston for transportation. Do you think that would be OK? Helping set up the garden?"

A pause, a terrifying silence, ensued. Then clouds parted and the sun broke through.

"I'd like that, very much in fact. As long as it's just the two of us."

William was confused. He didn't think he'd invited anyone else along. Saffron pointed skyward. "Just the two of us, at least for now," she repeated.

"Oh, OK."

They sealed their new understanding by pinkie swearing. One of Belle's little boys launched himself at Thor and did a half Gaynor off his side, startling the big fella into running all the way back into the barn. Saffron and William both laughed till they cried.

Mike Horvath had his own self-assigned task. He had to be careful not to step on the toes of the lead investigator for the murder, but he felt time was of the essence in a particular matter. After finding the second set of the same tire tracks, he called the forensics guy. He was informed that the tracks belonged a Pathfinder 17" All Terrain tire. Expensive and not real common, but still somewhat popular throughout Colorado. It wasn't much to go on.

He had done a little quiet digging into the character of Alex O'Connor and found he was an unsavory piece of work. Numerous investigations of misconduct, and warnings in his files. He also found out where he parked and what his schedule was for the next 24 hours. After his morning oatmeal tomorrow, sans Ex-lax, he planned on racing to Grand Junction and checking out the employee vehicles and then racing back in time for his own law enforcement shift.

George was anxious to relax with his gal for a bit. The two of them poured large glasses of Beaujolais and sat on the patio. From there they could see the two young people enjoying the young goats. The sunset was going to be epic. Mike, having completed his security duties, came down from the parking lot.

"Hey guys, there's a fender bender down by Hotkiss that I need to work. George, are you going to be here a couple of hours? Excellent. I will tell Danny Shapiro on the sheriff's night shift not to come till around 7:00 PM. I'm gonna head home and get some sleep after the collision call out, because my day will start very early. Be cautious and be observant."

After he was gone, some serious snogging was indulged in. For the first time in their courtship, Laurelyn allowed George's hand to rest in some interesting places. Far from being the whore of Babylon, Laurelyn had been very strict on the go-slow approach. George had not shared this with his son, figuring it was none of his damn business one way or the other, and he could keep his self-righteous judgments to himself. They watched in contentment as William and Saffron shut all the goats back in the barn, and then went to close in the turkeys and chickens.

As the young people approached, George withdrew his hand into his lap.

"Mind if we join you?" William inquired.

"Not at all, please do."

William grabbed a Hansen's soda for himself and Saffron helped herself to the rest of a bottle of Mesa Splendor white, poured over ice cubes and with a dash of honey stirred in.

"Silly Saffy," George ribbed, watching her trying to stir the honey into the cold drink. They all watched Beth close and lock both greenhouses and head toward them. She joined them for half a glass of Beaujolais and then headed home to rest and prepare for her next day's subterfuge.

Eventually William and Saffron went indoors to light the woodstove and chat in the warmth. Outside George and Laurelyn created their own warmth.

Chapter 14

The next day was cold and clear. Laurelyn made a heap of French toast with some of the eggs Saffron had gathered yesterday. Then they conferred about the day's plans.

"I will be gone from 8:45 until 10:00 or so. If you can milk Belle a little, that would be great. Just a couple of cups worth would be perfect. Then all of the regular stuff can be done with Weston's help. He should be here around 9:00. George and William will be unavailable but Sandy should be able to help if need be by 10:00. Weston will want to finish plowing but if he gets all that done, there's five pounds of ornamental flower seeds that Beth ordered to go in your garden. I'm not sure where, but I bet you know. Those can be scattered and raked in. A light covering of old shavings will help to keep the birds from devouring all of them."

That was a full day's work for them right there. Presumably, Laurelyn and Beth would both be there by lunchtime at the latest.

Laurelyn finished breakfast and dressed in the only business suit she had to her name. Makeup and hair done, she barely recognized herself. Saffron stared open-mouthed. She hadn't' been this impressed with a woman's appearance since Bernie had shocked everyone with her outrageous undercover get-up last year. Laurelyn did a tiny runway spin and said to Saffron, "You think this will do to impress a banker?" Saffron could do nothing more than shake her head up and down vigorously.

As Laurelyn and the crew headed toward their respective interviewees, Mike Horvath cruised the employee parking lot at the Grand Junction police department. To his disgust, and then rage, he found what he was looking for.

Laurelyn waited for Clive Lancaster to invite her into his office to set up a business account. Personal accounts he left to staff, but business accounts he always attended to himself. The civilities dispensed with, and the checks selected, the digital account was then set up.

"Remote banking is so critical these days," she said to Clive. "I find it so hard to get away from my place without something going wrong while I'm gone. Seems like there's always an emergency, even when there aren't dead bodies to cope with. Speaking of which, did you know the deceased very well? It seems like she was a young woman who grew up right here. Ach, poor thing, so young, to end up that way. Well of course no one should end up that way. I mean, can you imagine, being strangled and dumped into a toxic waste pit? Shouldn't happen to one's worst enemy in my opinion. Poor thing, everyone says she was a real beauty."

Laurelyn had strung together this stream of prattle, all the while watching Clive's face. The man was plainly distracted and becoming more so, and nervous, with each provocative sentence. Laurelyn was doing a superb job channeling Margaret Stanley. Seeing his facial response, she spun it a bit further, before going in for the kill.

“Yes, they say she was really popular in high school, homecoming queen, and all sorts of other things. Married well too, that nice accountant right here in town. And very civically minded, participating in all sorts of charitable works. Seems like she was a real saint.” The parry, and then the thrust. “Who would do something like that to her, someone so sweet, so full of life and promise? Such a loss to the community. I just can’t fathom it.” Direct, unyielding eye contact and then the twist of the knife. “I mean really, who would do such a thing?”

Clive’s expression had ratcheted from distracted, to nervous, to distracted, to angry and indignant, to focused and horrified. He rose from his seat, stiffly, and said loudly, “I’m terribly sorry, my phone alarm just vibrated to remind me of a doctor's appointment. I will have Jane in here shortly to finish the paperwork."

"That's fine and thank you. Just as long as I am set up with electronic deposits. I hope the service at this bank is excellent in that regard. Some banks' remote deposit services have trouble reading some people's signatures. It can be such a problem sometimes, that they mush the signatures together and just keep a standardized sample on hand to compare to, and then manually approve the deposit. Very cumbersome and inefficient, takes way too much time. Often the client doesn't even know that's the practice. I hope that is not the case here."

Clive stumbled over his own chair leg on his way out of his own bank. Cashiers and customers alike stared. His Acura screamed out of the parking lot.

Is it something I said? Laurelyn smiled to herself, abandoning the paperwork and sprinting after him.

Chapter 15

Sandy had concluded her assignment. She had also acquired a new accountant. Her probing came up with nothing more than a person saddened by past events and by his former partner's ugly end. He had even expanded on the topic a little, admitting theirs was not an ideal marriage, but that they'd had a few good times. He concurred that no one deserved to go that way and that he hoped they caught the bastard soon. Sandy, who felt that she and Mike were going to be needing help this summer, contracted with Jack to assist her with payroll and taxes. She would slog ahead with this year's but didn't want to have to do it again in the future. They shook hands and parted just as McGinty arrived back from Denver, needing to brief his client on developments. She felt certain Jack was innocent.

So even though no murdering spouse was cornered and forced to confess to evil deeds, at least she got a good accountant out of the deal. And for a discount, since some of Jack's clients had dropped him recently. The bank being one of them.

Beth found her mark to be obnoxious, arrogant, and probably capable of all sorts of sketchy legal shenanigans. He would never get any of her business in this lifetime, she concluded, but he just didn't seem to have what it took to be a killer. A schoolyard bully yes, but the type that went crying to teacher the minute an older kid stood up to him. No firm conclusion, and he should still be considered. But her instincts voted against him, both as a human being and as a killer.

George and William sat in the pastor's tastefully appointed office. They asked about youth involvement and activities. Bible classes and religious instruction were described. Charitable works that were supported by the young people were a great source of pride to the reverend, but of course, God deserved all the credit. Still, a multi-generational program, where the youth group sponsored reading programs for little kids, read to seniors who were homebound, and volunteered to help with GED instruction for local older teens who were unable to graduate from high school by the usual means, sounded pretty noble. His passion for education impressed George, and he felt some of his own rigid attitudes and preconceived ideas softening. He used the topic of high school students to mention the demise of one of Peony Creek's own. The reverend paled and then had a sudden coughing fit.

"Sorry about that. I had a visit from a detective yesterday. Very intense young woman, as I guess you would have to be. The interview left me shaken, and I am still processing it through prayer and meditation. A terrible tragedy she thought I might know something about. I don't, though." His eyes slid away from George's.

Liar, thought George. *Doesn't God frown upon that*? George felt his heart hardening against the man again.

"Surely, pastor, in a town this small, there must be some gossip?"

"Daaad!" William reproved.

"Just sayin, since it's a small world around here. A decent education may be something the killer lacked, and that or drugs drove him to do this despicable thing.

No clues, no guesses, none at all?" George did not expect an answer from a man of the cloth and was just gauging reactions. He was lying, that was certain. But his physical tells weren't strong enough to indicate personal guilt. At least that was George's take.

"I just want to make sure that my son here doesn't get mixed up with the wrong kind. He's had his own challenges with drugs, so I'm a little protective."

"Daaaaaaaad!"

Duckworth decompressed immediately in a way that's hard to fake. "Of course, sir, and you are right to be careful. Responsible parenting. We have a no drugs policy, no marijuana, tobacco, or alcohol allowed on any outings or sponsored activities. Those who cannot abide by that, are encouraged to attend services but may find certain unsupervised activities closed to them. We want to give them as much help and support as we can, without encouraging the spread of any addiction. Righteous living in God's grace is enough of a high, I find."

George, knowing his son would turn 21 next month, really didn't have a leg to stand on. He felt ambivalent about the pastor and the whole church thing but decided to give the benefit of the doubt. He assured the pastor that William would be seen on Sunday mornings soon. With that, he took their leave. On their way out, William insisted on a quick tour of the sanctuary. It was lovely indeed and very modern for such a small town. While William ooohed and awed and bowed his head in prayer with the pastor, George fired off a quick text to Sandy. *'Hiding something. Might be our guy, but I doubt it.'*

Sandy, receiving a second ambivalent conclusion, drummed her nails on Laurelyn's table. She had stopped by to do a quick check and provide any support Saffron might need. She also wanted to hire Saffron for a couple of afternoons this week for some pruning tasks, since her farm chores too were behind. Saffron eagerly agreed and mentioned she might also consider William if she could pry him away from George. Both young people needed their own cell phones and were saving up. William would have been content with a burner, but Saffron, being environmentally minded, wanted something good that would last and last. Sandy thought it might be a good idea, and told her she would consider it.

She knew she wouldn't hear from Mike at all. Only dinner chat would pry his findings from those tight lips.

Fairly tight lips. So now, only Laurelyn needed to check in with her results. Just as she finished thinking those thinks, the awaited text arrived.

'It's the banker.'

Chapter 16

Laurelyn hopped into her own vehicle and followed him. Clive headed straight for his home, a stunning log mansion even bigger than Beth's place perched on the edge of one of the mesas. The security gate was a little frozen at that elevation, and although it began to close behind Clive's car, it did so at its leisure. Laurelyn, not being able to believe what she was doing, was able to park outside in the woods and slip through the gates just as they groaned shut.

This was Bernie's gig, not her's! What in the name of almighty God had seized her? Excellent question to have asked earlier, since now she was trapped. Guard dog? Better hope not.

Shaking with anxiety at her foolishness, she crept up to the house. A tiny little bark sounded from the inside. No Rottweiler, that one. Thank God.

Now she was the lurker, and like the other one, was drawn to a spectacle in silhouette unfolding. Clive was pacing the room, back and forth and back again. There appeared to be no one with him. But she heard his raised voice, just barely audible through the thick walls. No words, just tone.

She saw what appeared to be Clive with a load of clothes in his arms that went upward and then down. Down into a suitcase, perhaps? He disappeared for a bit, then reappeared, but still bending down. Was that the sound of sobbing? His? He disappeared again. This time pounding steps as he descended the winding, heavily ornamented wooden staircase were plainly audible. Laurelyn sank behind the stocky ranunculus bushes adorning the outer walls of the home. Clive, suitcase in hand, raced for his car.

Throwing suitcase inside, he then spun the car around and headed back to his gates. He reached them long before Laurelyn, having a driveway nearly a ¼ mile in length. Her fancy pumps, worn for the occasion of towny spycraft, were nearly her undoing, sending her sprawling down the side of the driveway. Road rash stripped both knees of their skin, but she knew she had to reach the elderly gates before they closed. She did, but not before leaving a piece of the only good blazer she owned stuck between the latch and its receptacle.

Jerking her own car around and down the cliffside road, she nearly lost him. She reasoned he would not go up the valley toward McClure pass, but down to a place where highways and airports would speed his departure. She wasn't wrong, but once on an abandoned stretch of country highway, he hit 90, a pace her truck could not match, at least not for long. She was able to keep him in sight only long enough to see that up ahead he paused at the highway intersection in Delta and turned north. Grand Junction. More flights, more times of the day to more places than Montrose. Also, a tributary to Highway 70.

Would he flee the area via car, or the country via international flights from Denver? Impossible to know. She pulled over and called Sandy.

"Clive's making a run for it. He's headed for Grand Junction."

"Mike is headed back from there right now. I wonder if he can intercept. Probable cause? Speeding?"

"No probable cause yet, but speeding, absolutely. The vanity plates read, "PC BANKER." Cobalt blue Acura, very new. He only took one suitcase, so I bet he's going to the airport."

"Suitcase, you know this how?"

"No time, get Mike on the job before the killer gets away."

"Yes, Inspector Frost, immediately."

Mike called out the door for another box of Kleenex. The man simply would not stop crying! All Clive Lancaster would say in between bouts was, "She shouldn't have to pay, not her, not pay…" and then more tears.

Clive had been pulled over going 120 mph. The trooper that pulled him over had already written out the citation when a call came in on his radio. He looked at the plates and called it in. When he returned to the vehicle hand on gun and said, “Sir, I need you to step out of the vehicle,” Clive had burst into tears and hadn’t stopped since. Clive was driven back to Delta, to an interview room, where he now sat with Mike. Mike was waiting impatiently for the detective, who had been detained by a mudslide caused by the unusually warm spring weather. It seemed like half of one of the mesas had come down, bringing a barn and several old shacks with it. She had been in Peony Creek and now would need to go back up the valley to Colbran, over to Hwy 70, and then south on Hwy 50 in order to get here, a detour of at least three hours. In the meantime, Mike had been tagged. There was little point in trying to interview this suspect though, in fact, a sedative might be in order. But he gave it the old college try.

"Who shouldn't have to pay sir? Tammy Splann? Do you know the deceased? We have testimony that you did and that the two of you..." the wails that issued forth from the man would have brought down the walls of Jericho.

"Love of my life, she deserved better, what kind of God..." and the weeping resumed.

Feeling utterly out of his depth, and wishing Nancy Clifton were here instead of him, he simply did the only thing he could think of and sat down next to the handcuffed man. He sat in silence with him, bearing witness to what must be indescribable guilt and anguish.

Time passed, and still no detective. But at some point, the silent vigil seemed to sink into the prisoner's brain. His tear-swollen face emerged from his arms. He looked at Mike, and to the extent that he could, placed his hand on Mike's forearm.

"My wife, someone needs to take care of my wife. She is innocent in all of this. Wanted to be with her, with her…at the end."

The moment of clarity gone, the tears resumed. Now Mike had warrant to call Nancy, thank God. He texted her for a '*welfare check on Sharon Lancaster. Urgent. Details later. EMT may be needed.*'

The Peony Creek investigative squad were all on the east side of the mudslide. Mike had informed Sandy that he was stuck in Delta with a suspect and likely would not be home before midnight if he had to go over Grand

Mesa to get home. It turned out that Nancy was on the west side of the slide, while the Lancaster home was on the east side. Nancy relayed this info to Laurelyn via voicemail, since she didn't pick up.

Returning home from her hot pursuit toward Delta, she had been climbing the grade toward home, when just past Hotkiss, she heard a rumbling. Her aching, lacerated knees and general emotional condition could not process the sounds. Not a flat tire!! Pulling over and getting out, she started to check for a flat when the sound grew louder. *Can't be me,* her stress fuzzy brain chuckled with relief. *I'm stopped.* But then she saw a substantial portion of a mesa just behind her give way, roaring downward.

The lateral ditch had been filling slowly, with snowmelt slumping into the irrigation ditch canal that fed 352 farms along its length. The ditch was technically open on April 1st, but it already had a fair amount of water in it in advance of that first official day, three days hence. The sudden added snowmelt was too much. A section filled with water, then mud, and then with a groan, like those of the Lancaster gates, gave up the ghost. The embankment crumbled, moaning and then screaming under the weight of rock, mud, and water. Laurelyn, not sure she was safe, crammed her blood-covered, stiffening knees into her farm truck and took off.

Nancy responded to Mike that she couldn't get there. Mike called his wife, who called Laurelyn. Laurelyn couldn't help, she didn't have a control for the gate anyway. She called Nancy, who mentioned that some hospice workers had access, and Sally Handy did as well. Sally? Really?

Laurelyn called Sally. The New Age greeting grated on Laurelyn as much as her knees did at that moment. She ignored both and asked, without preamble, "Sally can you get into the Lancaster house?"

"Well hello to you too, and why do you want to know?" New age only went so far with Sally. "Sharon Lancaster is in danger, right now. Can you get up there?"

"The Goddess Is My Wings." With that, she hung up. Laurelyn only hoped that was a yes.

Sharon Lancaster's nearly blind eyes gazed at Sally with a kind yet puzzled expression. She was a sweet soul but no fool, and her husband's behavior and departure told her something was badly wrong. Sally sat by her side, holding her hand. "Clive has done something, something wrong, hasn't he?"

Sally answered truthfully, "I really don't know. But I'm here and we will get through this together."

"Enough of that silly business, dear, there is no 'together'. Not that I don't love you, I do, and you have been a blessing. But I am crossing over, and I heartily wish for you not to follow me until your time comes. Is that understood?"

Sally smiled at this ferocious yet noble soul that she had come to respect so very, very much. She had inherited this person's herbal formulations from Sonora Tillman, the former owner of the Merry Clary.

"As you wish, my lady," knowing one of Sharon's favorite movies was, "The Princess Bride." Sharon smiled, then squinted at Sally. "You look drawn, dear, are you getting enough sleep?"

This was dangerous territory because Sally had been sleeping little lately. Last week, some late-night frolicking with Weston had robbed her of somnolence, this week, the news of bodies and investigations had troubled her. She had wondered if giving the crew at Bad Aspen her yearbook had been the right move. Given the haphazard scuttlebutt from Weston, it might have helped to catch a killer. But at what cost? This dear woman whose hand she held had less than two weeks to shuffle off this mortal coil, so her doctors said a week ago. Sally wished for nothing more than for this person to depart in peace. That seemed to be threatened right now.

"Did I tell you how Clive and I first met? Oh, I probably have but indulge a dying woman. He had just earned his junior varsity lettermen's jacket. He was a good athlete, but some of the boys bullied him because he came from money. Nonsense, good performance is good performance. I was walking home with friends and heard someone sobbing under the bleachers. The other boys had been hard on him, saying he'd bought the coach off. My friends, seeing who it was, said, 'Oh it's just that rich pussy having a moment. Ignore him." But I couldn't, you see. I had never heard him say a mean word to anyone or use his family's wealth to impress others. Quite the opposite, he seemed to be ashamed of the money. And so, he excelled at sports, he truly did, because he simply loved it. Even weightlifting, eventually. The other boys were just jealous. Anyway, I asked him what the matter was, and he looked up and said, 'Why do you care? And I, being a teen, said, 'Just cuz.' Three years later, we were married. Ah, that man, he had his weaknesses, sure. I think my illness may have caused him to stray. But the good times we had, the love, and laughs…"

Sally, imagining those days, felt images of those by-gone days filling her mind's eye. She later thought that maybe Sharon had whispered, "He's not coming back, is he?" But it would have been so soft, she could have imagined it. What Sally did remember, later, that when she began to ask Sharon about their wedding, the hand in hers was cool and limp. Sharon had crossed over.

Sally called Nancy, who called Mike. Mike asked her to come to the Delta Police Station. That she could do, being west of the mudslide. She entered the interview room. She signaled Mike to leave, which he was more than happy to do.

"Mister Lancaster, I have some very sad news for you. I am so sorry to have to tell you this at this stressful time, but your wife Sharon has passed away."

To the end of her days, Nancy Clifton would tell co-workers that never in her life did such sad news bring such an unexpected response. Clive, hearing the news, still weepy but seeming to be more present, asked her to repeat it. She did, confused and concerned. Still, this was not an entirely unheard-of reaction. Next would probably be denial, or a claim that she was lying, she had to be. Instead, Clive wiped his eyes. He became completely calm and said, "Please bring in Officer Horvath, I wish to make a statement." Stunned, Nancy did so, after asking if he was sure.

"I've never been more sure of anything in my life."

Over the next two hours, Clive dictated how his wife's illness had affected him, and how, like the coward he was, he had reacted. Then, as she got worse, he had received a letter from his affair partner, asking for $3,000 per month in blackmail money. Otherwise, Tammy would expose his infidelity to a wonderful, delightful, kind soul, dying because God had no mercy. He had paid the money, but there had been hints that it was not enough and that she might just let it leak to the bank's charities that something deliciously scandalous was happening. Maybe $4,000 would be better.

It was at that point that Clive knew that the money was not nearly as important as the influence, the ambition, and the power to her. So, he decided to end it, the only way he could think of. He sent her an emerald necklace. He met her for one last assignation. And then he strangled her. He had enjoyed choking her in previous trysts. It was so very much something his wife would not have gone for. This time it was for real. He had every intention of killing her, let the record show.

But once the deed was done, he knew he had to throw potential investigators and her husband off the scent. So, using the check franking service the bank employed, he convinced a clerk to create the notes. He convinced her it was for a mystery dinner theater fundraiser and they would be used as clues.

"The bitch would have gladly stolen that wonderful woman's last peaceful moments on earth. She didn't deserve that, she didn't deserve any of this. But now she's gone, and beyond the torments of Tammy, me, MS, or anything else. I did it, and I deserve to be punished. No more hiding, no more lying. God, what a relief!"

An extraordinary confession if ever there was one, at least in this valley.

"And the note on the sharpened stake," Mike asked with more hostility than was professional.

"Oh that, that was O'Connor. He was the only creep I knew Tammy had also blackmailed that I could call to help me get her body out of the hotel room. I hadn't thought that through, I just wanted her dead before she could contact Sharon. He came over, and together we wrapped her favorite blue erotic strangulation scarf around her neck and dragged her upright between us, out of the room, and down the stairs. We talked to her like she was a drunk in case someone was watching. No one noticed us. What became of her after she went into Alex's SUV, I neither knew nor cared. We never spoke of it. I paid him $10,000 to help with his divorce and that was that. I guess she ended up in a pit on that farm up the road, the one run by that nosy bitch that posed as a client this morning. May she and Tammy both rot in hell. O'Connor had been keeping a watch on the pit and saw her out there over and over. And not just her, but the cop's wife too. Oh, I guess that would be your wife. Sure, I was worried, and let him know it. But that note, and that stake, that's all on him. That farmer lady may be a nosy troublemaker, but I would never randomly hurt an innocent person. That would be immoral. O'Connor would though. He's a rattlesnake to the core."

"OK, I think we've heard just about enough for now." Mike turned the recording equipment off. Clive, with his dignity restored, pushed his large, cuffed hands through his graying hair and performed the most dignified perp walk that Mike had ever seen. An hour later, Alex O'Connor was arrested on a huge number of counts, one of which was accessory to murder after the fact. And now Mike and the Peony Creek investigation team could finally rest in peace. In a good way.

Chapter 17

Easter Sunday dawned as if the world itself were being resurrected and renewed. An iridescent cloud display over the West Elks made anyone with a soul stop and stare. The Sunday shutterbugs lined the roadway where Alex O'Connor had so recently parked. The arrest of both Clive and Alex had made the fencing around the pit no longer necessary. The field had been plowed again and then seeded yesterday. Today, Laurelyn had declared a day of rest. The basics would be done, of course. No animals would be neglected. But anything that could wait, would wait, until tomorrow. For today, it was Easter Egg hunting.

Saffron, Eddel, and Tiffany were in the kitchen dyeing eggs. Once Sunday services were out, William and George would join Laurelyn and Beth, the Horvaths, the Riveras, Weston and Sally, Milton and company, Saffron, and Nancy Clifton. Oh, and Deborah Johnson.

To Laurelyn's surprise, that had been Saffron's idea.

"If you're up for it, I think it's brilliant. She really is trying. I think this is a good, neutral way to begin a reconciliation. If that's what you want."

"I'm not sure, but I'm willing to be open-minded, a little." When Saffron had returned her mother's call yesterday, Deborah had burst into tears of joy.

Every refrigerator and cooler in the house and the milking room was crammed with foods for the post-egg hunting barbeque. Dorcas was supplying the sauce, an American/Ethiopian custom concoction. Weston and

George were supplying the beef. The Horvaths supplied the wine. Sally was bringing a huge crock of custom herbal iced tea, with real hibiscus. The Riveras brought homemade coleslaw, and Nancy brought store-bought potato salad. Nancy didn't cook much. Beth supplied the beer. Deborah Johnson was on a shoestring budget, so Laurelyn suggested that her contribution could be clean-up.

By 10:00 AM, all the eggs had been dyed. Beth took the little ones into the herb and veggie greenhouse so they wouldn't see where the adults were hiding five dozen eggs. The entire property, except the greenhouses, was fair game. All adults except Laurelyn and Beth hid the eggs. Deborah's radiant face was payment enough for including her, Laurelyn concluded. The chickens had been released for the day, to have access to fresh greens, and the turkeys, Tom and Jerry, were out to provide comic relief.

George and William arrived. They had to park on the road, the farm's parking lot was so crammed. William went immediately to help Saffron hide eggs. Laurelyn pulled George aside.

"So how was church?"

"Not as bad as I'd expected. Duckworth gives a mean sermon, I will admit. The theme was interesting. Not so much about the resurrection, as about scapegoating. With Jesus as the principal goat. Unusual but it seemed to be well received. By everyone but the detective on the Tammy Splann case. She was in the congregation. She stayed behind after everyone had left the sanctuary and then ushered the reverend out the back door. I suspect she had a squad car waiting to take the reverend down to Delta for a far more difficult grilling than the one she gave him before. He was indeed hiding something, the identities of the other blackmail victims, all of whom were potential murderers, including himself. At least that's what Margaret Stanley whispered to us as we were leaving. The detective asked him directly four days ago if he knew of any reason certain persons might be involved in Tammy's murder, or anyone he thought might have an interest in framing him, and he denied any knowledge. Now he's liable for obstruction of justice charges. A shame, really. He's worked so hard to build this congregation and do good works in the community. I found I rather liked the man, but now he may be run out of town by the more righteous folks among us. Only time will tell. Now if you'll excuse me, I have eggs to hide."

Eggs hidden, Beth, Tiffany, and Eddel pelted from the greenhouse. Beth was an honorary child since she didn't help hide eggs. She was careful not to find too many, and to utter squeals of delight whenever she found one. After a frantic 40-minute search, the girls had only recovered four dozen eggs. The last dozen were sleuthed out by adult/child teams. By lunchtime, they had found all but two.

Laurelyn had gotten the coals going and they were perfect, awaiting the brisket that would go on, roasting over the mesquite wood chips that Beth had special ordered. Plastic wine glasses came out, and the beer cooler got opened up. The beautiful weather allowed the guests to take off their coats and hang out in T-shirts, as long as they stayed in the sun.

"Hello, the farm!! We come bearing gifts." Laurelyn looked up to see two uninvited guests walking down the path. Uninvited, but not unwelcome, since they carried a case of decent champagne between them. Jack Splann and McGinty Faraday had no close relatives in the area to spend the day with. McGinty, wishing to avoid holiday traffic over the pass, elected to stay and celebrate his easiest murder defense to date. His only murder defense to date. He still charged Jack an arm and a leg for that privilege, so figured a case of champagne was the least he could do. His arrival earned him the stink eye from Sally, but since the day was both bright and sacred, she decided to tolerate him.

With a pointed look, she took four bottles from the case and went to the milking room, where she knew there might a little extra fridge space, the better to keep the bubbly cold. Unabashed, McGinty swept her a courtly bow, and then popped the top on the first bottle. The cap landed several yards away in the vicinity of Tom and Jerry. Tom, thinking it must be a really large, fat, flying insect grabbed it. Jerry, eager to take it from him waddled in hot pursuit. The resulting hilarious scene caused more bottles to be popped in that direction.

"A toast! To exonerating the innocent and persecuting the guilty. To justice served!" All drank to McGinty's toast.

"And to the Peony Creek Sleuthing Society!" Sandy declared. Everyone but Mike toasted to that one.

Laurelyn did not hear her phone ring but felt the vibration. Seeing the name, she shushed everyone, which took six more rings to achieve.

"Hola, mi amiga! Mi madre say you bean calling me. Que paso?"

"Oh Bernie, at last. There's been another murder."

"What, no shit!" she responded, dropping the fake accent. "I'm going to start calling that place Death Valley. Do you need my help?" There was a long silence while Laurelyn switched her phone to video. Then, sweeping it around the assembled throng, she pronounced, "Not anymore. We muddled through, and a killer was brought to justice. Now we are celebrating. Wish you were here!!"

"Ay caramba! You kids have all the fun. When I retire, I'm moving out there. Never a dull moment, and always a dead body. I want details, ASAP. First thing tomorrow if your head is not too sore. Or too big." Laurelyn stuck her tongue at her and signed off.

"So that was the infamous Bernie Gonsalves, of whom I have heard so much," McGinty observed.

"All true stories." Laurelyn, Beth, and Sandy all said at once. Beth was the first to cry, "Jinx."

"Not a chance," Sandy exclaimed.

After a couple of hours, the adults were as well basted as the beef. They indulged in nostalgia and made numerous plans for the coming spring and summer. The pandemic had decimated the travel industry, so Sandy and Mike were planning to expand the vineyard and wait it out. William and Saffron agreed to work one afternoon each week for cash. Just when folks started to settle into their post-lunch comatose states, George hollered, "Not so fast. There's gingerbread to tackle." Everyone but William thought he meant dessert, and they groaned.

"Forward troops, there's decorations to install. We march forth to Saffron's Garden!" William had popped up to their vehicle and grabbed eight wooden scrolls and the brackets to mount them with for the gazebo. Handing out hammers to the men, George marched at the front of the parade like a drum majorette, his own hammer pumping up and down ahead of him.

"Boys." Laurelyn shook her head, grinning. She started to carry plates in, when Deborah said, "Now you stop that, this is my job. Go supervise those fellas so they don't' hurt themselves. Or each other." Grateful, Laurelyn set off to do just that. Weston already had ladders set up, so a flurry of hammery sounds issued forth. After watching for a bit, and enjoying the delight on Saffron's face, she thought she heard another sound.

Hard to be sure. Probably nothing. Wait, there it was again. GOAT!! The last doe must be presenting.

Laurelyn quietly mustered her troops. Beth, Sandy, Dorcas, and Cindy Rivera all went to assist the new arrivals. One baby was already on the ground. Beth grabbed the towels and the birthing kit. She and Sandy began to rub the baby down. Cindy, not knowing what to do, stared until Sandy held the clean baby up and commanded, "Dip!" She tipped with her head to the birthing kit just out of reach. Cindy, still confused, held up a jar of KY Jelly.

"No, the dark red stuff in the other container. Now put the baby's umbilicus in the bottle. Shake it up and down so it splashes up onto the tummy a couple of times. Great, now hold this little, little…GIRL, WoooHoooo… up by her mother's face."

Another set of hooves, with a tiny muzzle on top. Not a breech! Laurelyn was able to pull the baby right onto the towels, the way she preferred. Nose and ear cleaning, rubbing, and dipping, all done in less than six minutes. Another groan. Triplets again! She hadn't seemed that big. Two pushes, some blood, and the two more tiny feet and a face. Another correct presentation. This indeed was a happy day. The third tiny baby slid out easily.

"Beth, we got this. Can you make oat broth for the mother?" Beth was away in a flash and missed the glorious announcement.

"GIRLS, ALL GIRLS." This seldom happened, and this was the first set of triplets born at the farm where not one was a male. And their colors were stunning. They had all taken after their brightly spotted father, and not the elegant, but more plainly colored, dam. Sally, late to the party, was in time for the jubilant decree and grabbed another bottle of bubbly from the nearby milking room fridge. These girls, once registered and sold, would pay for all of the seed ordered for the new pasture, and then some. That is if Laurelyn could part with them. She thrust those unfarmerly thoughts away. Time enough for that later. This birth had been so blessedly easy that the men were still hammering away and the womenfolk hadn't even been missed. It was Weston's sharp eye from atop his latter that broke the spell. He saw Beth hurrying from the house with a steaming stainless steel soup pot filled with oat broth. That could only mean one thing.

"Hey guys, I think there's new baby goats. Right now." Then he kept hammering. Simon, Milton, Adolfo,

Jack, and McGinty, all broke for the barn. George, knowing that he wasn't needed and that there were already too many onlookers, stayed put.

Sally leaned against a wall. Tears came to her eyes. She asked Laurelyn if she could sit beside the mother and hold the littlest girl, the last one born. She wrapped her in a towel, and rubbed, rubbed, rubbed her. She pressed the little one to her heart and whispered to her. Laurelyn noticed the emotions happening and gazed at her kindly. Sally looked up, and croaked, "Sharon, can you name this one Sharon?"

"Of course, from now on, she is Bad Aspen's Saintly Sharon. That will work just fine. Her mother is 'Saint George's Queen Anne.'"

The little girl, warm and strong enough to wiggle, and being held too close for comfort, gave a loud, "BAAAP."

"She approves!" Beth knew that was not the case, the baby just wanted down, but felt it would lighten the mood, which it did. Sally smiled a fragile smile and put the little one in front of her mother, where she promptly collapsed. But not for long. Within a few more seconds, she was on her trembling legs.

"A fighter, that one, like her namesake," George spoke from behind that crowd. He had plastic cups in hand, and the last of the champagne.

"A toast, to those who fight the good fight, both near and far. To Sharon!"

And that was a sentiment everyone was anxious to toast.

Epilogue

The Easter crowd had thinned significantly. The only people left that didn't live at Bad Aspen were George, William, Weston, and Sally. They sat on the patio sipping Irish coffees to stave off the cold. The sunset was just as entrancing as the dawn had been. Weston and Sally sat smooshed together under polar fleece blankets. Laurelyn and George huddled in a likewise fashion, with Laurelyn's head on his shoulder. William and Saffron sat together, but not under the same blanket. Too soon for that.

George, not knowing he was observed, gazed down at Laurelyn's weary head. Her hand drooped, and he tenderly took the mug from her before it could crash to the flagstones. Then with that same hand, he caressed the top of her head.

Sally, having once been George's lover, noticed these tender motions. She felt sad for a moment, for George had never been that doting with her, even during their most intimate moments. Overcoming a twinge of self-pity, she caught Weston's eye and nodded her head in their direction. She then whispered in his ear. Weston nodded in return. They quietly finished their own mugs and rose silently. Putting fingers to lips, they motioned the other two young people to rise silently as well. The four went into the house.

"It's time," Sally pronounced. "Time to leave these two alone. William, we'll give you a lift home."

"But what about my dad, how's he gonna get home?"

"Let the two of them worry about that in the morning, shall we? They've waited a long time, and tonight they need each other. And I know you know what I mean."

William's eyes widened. "You mean they've never…?"

"That's what I mean. Don't ask me how, a woman just knows these things. Laurelyn's had a hard few days, hell, a hard few months, and could use some comfort. Come on now, kiddo. No arguing."

William let himself be shanghaied away home. He felt like an idiot and spent the next hour at home in prayer. He plainly had a lot to learn about adult life, and all its complexities, never mind religion.

Saffron gave George's shoulder a quick squeeze and departed for her own room, shutting the door behind her.

After the sun had fully set, and a serious chill took hold, he picked up his love and carried her to her bed, where he undressed except his underwear. He slipped in beside her and enfolded her in his arms. He could feel passion stirring but fought it down. This night was hers, and up to her.

Before starting in the greenhouse the next day, Beth had given the silent George a lift home. She didn't grill him although she dearly wanted to. He thanked her politely at his doorstep and then went inside. He looked around for William or Carlos. He spied them both in the vicinity of the stable. Alone in the house, then.

"YEEEEEEESSSSS, yes, yes, yes, yes, yes, yes!!" He couldn't stop himself. Laurelyn had awoken him with her hand on an interesting and interested part of his anatomy. With no further ado, he accepted the invitation and began to kiss her all over her body. But before their final merge, he wanted to be sure she knew how he felt.

"I love you, you know that don't you?"

She gave his exposed shoulder a tiny cuff. "Of course, I do, you big galoot. And I love you too."

"Big galoot!! Is that the best you got!" He wrestled with her a little, his knees grazing her mangled ones.

"Oh, oww, watch the knees!"

"Oh God, sorry, sorry, sor…"

She stifled his apology with her lips and made sure he got no sleep the rest of the night.

After George's morning departure, Laurelyn hummed a little here and there, smiled randomly here and there, and generally made it hard for Saffron to pretend she didn't notice. She finally departed for less embarrassing pastures.

All 14 baby goats were doing extremely well, and she wanted to make sure they stayed that way. As she supervised baby goats in the big paddock, she watched Laurelyn sashay down to the chiltibarn. She had never been in that kind of love and wondered what it must feel like. Maybe one day.

In the chiltibarn, Laurelyn's phone rang.

"You promised, mijita, to spill the frijoles. And now look at the time. 10:00 your time! You call that a farmer's morning? What happened to 5:00 AM? I guess that means that party was a success."

"Uh-huh," a musical lilt with a slight rise on the end.

Bernie, who was known for her uncanny powers of observation, said, "Oh, OHHHHHH! That kind of success. It's about time."

"You would say that, wouldn't you. The time was just right, exactly as it should be."

"Don't you sound smug! Details on that later, but now, for the murder-solving bit."

"Well, it started when Weston was plowing the new pasture, you know, those adjoining acres that Beth bought for us, when he spotted, oh hold on…Sandy's texting and she says urgent. '*Alex O'Connor hanged in jail cell. Mike suspected.*' Holy crap Bernie! Not Constable Mike!"

"I'll be there tomorrow. Got a lot of leave coming and can't think of a better way to spend it."

Disconnecting, Laurelyn thought to herself, *I can!*

Here we go again.

For fun facts about the author and to get background on the real Peony Creek and even some of the recipes from this book, why heck, just mosey on over to the website http://www.laurenhallruddell.com

Made in United States
Troutdale, OR
11/03/2026

56844429R00193